Retain - Sept '08.
BR - 06/11

I/P
No later ed.

WARWICKSHIRE

KV-294-233

007165

WITHDRAWN

93
94
96

WITHDRAWN LIBRARY

Warwickshire College

00531981

METALS FOR ENGINEERING CRAFTSMEN

LIBRARY
WITHDRAWN
WARWICKSHIRE
COLLEGE

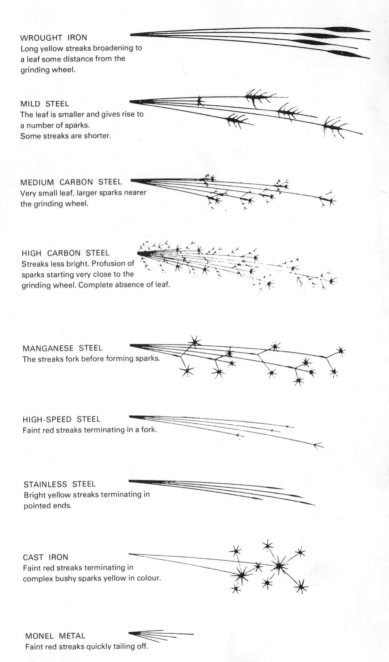

WROUGHT IRON
Long yellow streaks broadening to
a leaf some distance from the
grinding wheel.

MILD STEEL
The leaf is smaller and gives rise to
a number of sparks.
Some streaks are shorter.

MEDIUM CARBON STEEL
Very small leaf, larger sparks nearer
the grinding wheel.

HIGH CARBON STEEL
Streaks less bright. Profusion of
sparks starting very close to the
grinding wheel. Complete absence of leaf.

MANGANESE STEEL
The streaks fork before forming sparks.

HIGH-SPEED STEEL
Faint red streaks terminating in a fork.

STAINLESS STEEL
Bright yellow streaks terminating in
pointed ends.

CAST IRON
Faint red streaks terminating in
complex bushy sparks yellow in colour.

MONEL METAL
Faint red streaks quickly tailing off.

THE SPARK TEST: We gratefully acknowledge permission to reproduce this diagram
from the 'Quasi-Arc Welding Manual' published by The British Oxygen Company Ltd.,
Hammersmith House, London W.6.

007165

METALS FOR ENGINEERING CRAFTSMEN

a guide to their composition,
properties,
and manipulation

In 1988 the Development
Commission merged with its
agency CoSIRA – the Council
for Small Industries in Rural
Areas – to form the Rural
Development Commission

LONDON
COUNCIL FOR SMALL INDUSTRIES
IN RURAL AREAS
1979

PUBLICATION 75
First published 1964
Revised and reprinted 1979
Registered Office:
© CoSIRA, Queens House, Fish Row,
Salisbury, Wilts SP1 1EX

ISBN 0 85407 027 3

Warwickshire College
Library
Moreton Morrell Centre

Class No: 669 ~~WITHDRAWN~~

Acc No: 00531981

Made and printed in Great Britain
by McCorquodale Printers Limited, London

CONTENTS

FOREWORD

Metallurgy is a highly specialised subject, introduction to which is confined to fearsome-looking volumes full of technical data shrouded in the mystique of the professional, so making them intelligible only to the professional.

This is a great pity, for the subject is an exciting one, ranging from geology, through mining, smelting, refining, research and manipulation, into the many everyday commodities that we see about us, and this tendency to confining these realms of knowledge to formidable and expensive text books, effectively bars many an enquiring mind from even the most elementary understanding of the subject.

Thus it may be difficult for many to appreciate that although metals have been in use for longer than recorded history, the entire development of making them in large quantities, and with such specific characteristics, has been confined to the last ninety or one hundred years, and it was during this period that such far-reaching and thrilling discoveries as manganese steels, stainless steels, the electrolytic processing of aluminium, beryllium copper alloys, the nimonic metals, and many others, were made.

Some were accidental; some just happened; but they were no less significant in their effects on industry. The nimonic alloys, on the other hand, were the result of intensive metallurgical and engineering research extending over the years from about 1936 as a result of a definite programme to produce alloys capable of retaining their properties at the very high operating temperatures of the gas turbine engine.

It is hoped that this book, which has been compiled for the guidance of the small users, will disperse some of the mystery that has in the past been responsible for 'blinding them with science', and whet their appetites for a greater understanding of the structure and properties of those materials that they are so adept at shaping, whether it be on the hearth or on a machine, but from which, and by their skill, they earn their daily bread, and contribute to the needs of the nation.

S. H. M. Battye, C.B., M.A.
Formerly Director, Rural Industries Bureau

ACKNOWLEDGEMENTS

Those who study this book and derive benefit from it will, we hope, pay tribute in thought to the late G. I. Wilkes, A.I.Prod.E., who gave so much of his time and experience to compiling this publication. To his name must be added that of Gordon Carter, M.S.I.A.D., who designed the book and supervised the printing, R. L. Aston, A.M.I.Mech.E., M.I.Prod.E., F.G.S., Head of Department of Production Engineering, Lanchester College of Technology, the technical editor, R. A. Ballard for revising the second edition, and Jeremy Towner, Department of Metallurgy and Materials Technology, University of Surrey, for providing the Introduction, to all of whom CoSIRA pays its grateful tribute.

INTRODUCTION

In the first edition of this book the composition, uses, properties and welding procedures for some 40 different metals and alloys were reviewed. In this revised edition some of the principles involved in the processing of metals and alloys have been added in the belief that this will contribute to a better understanding of the subject.

Metals and Alloys

All metals except gold appear in nature in the combined state, frequently as the oxide. Hence the reaction of metals with oxygen to form an oxide, i.e. 'rusting' or more generally 'corrosion', is the natural process. To obtain the metal from the oxide energy is required to reverse this natural process but, once obtained, the metal immediately starts to react with the environment to return to its natural state. Different metals corrode at different rates, but corrode they do—we can only hope to slow the rate down to an acceptable level.

The number of metals used in everyday life is very limited. However, the mixing of two or more metals to form an alloy increases the number of useful metallic materials considerably.

In practice the majority of metals and alloys start life as some form of casting, produced by pouring the molten material into an appropriate mould. Such castings will frequently have inhomogenous properties, due either to a variation in grain size and shape, and/or a variation in composition, through the casting. Such inhomogeneities are removed by annealing, i.e. heating the material to some temperature well below its melting point, the higher the temperature and/or the longer the time at temperature the more complete will be the removal of the inhomogeneities. The cooling rate from the annealing temperature will not influence these inhomogeneities, but may give rise to other effects which will influence the properties of the material.

Most of the craft industries work with sheet, bar, rod or wire material. These will invariably have been produced by casting into large rectangular moulds called ingots, annealed to produce a uniform grain size and shape and a uniform composition throughout the ingot, and then deformed to produce the required product. The production thus involves heat treatment, i.e. annealing, and mechanical working, i.e. deformation, both of which give rise to changes in structure and hence properties.

Behaviour of Pure Metals

Pure metals can be classified into two groups, one including copper, aluminium, and nickel, which undergo no phase changes on heating from room temperature to the melting point, the other, which includes iron, in which there is a phase change between room temperature and the melting point. The following discussion applies to all materials in the first group, and to materials in the second group up to the temperature at which a phase change occurs, i.e. up to 910 °C. in iron.

If an annealed material is bent gently and released it will spring back to its original shape, i.e. it has been elastically deformed. If it is bent more and released it will not return completely to its original shape, i.e. it has undergone a permanent change of shape by plastic deformation; some spring back will occur since a material always deforms elastically before it deforms plastically. If it is bent still further then fracture may occur. Thus on deformation, whether it be by bending, hammering, or some form of drawing, a material may respond elastically, plastically, or by fracture. Of these an understanding of plastic deformation is perhaps the most important and hence the following discussion concentrates on this.

On deformation the material is observed to get hot and to change shape. Further, as the deformation proceeds, more effort is required to produce a further deformation. Fig. 1 shows the influence of plastic deformation on the strength and hardness, measures of the work needed to produce the change of shape, and on the ductility, a measure of the amount of change of shape before cracking occurs.

It can be seen that the annealed material can undergo fairly extensive changes of shape with relatively little effort, whilst the heavily deformed material requires considerable effort to bring about a change of shape and the material is liable to crack. However, it should be noted that these changes in property occur

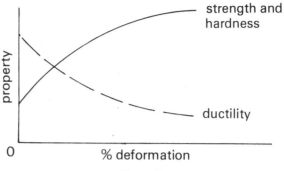

Figure 1

continuously with working, so as a material is worked it becomes progressively more difficult to work, i.e. it work hardens.

These changes in property on working reflect changes occurring within the material. These changes can be observed by polishing and etching the material and then viewing it under a microscope. Fig. 2a shows the structure of a typical metal in the cast and annealed condition, whilst Fig. 2b shows the effect of deformation on this structure.

The former shows uniform equiaxed grains—an analogous structure is seen by shaking up soapy water in a bottle—which becomes elongated in the direction of working on deformation; the markings, i.e. strain lines, indicate changes occurring in the material on an atomic level.

It will be recalled that the original casting contained inhomogeneities which were removed by annealing. In the same way the effects of deformation can be removed by annealing. Thus the work-hardened material can have its properties changed back to those of the original annealed material by annealing above a

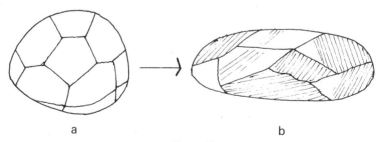

a   b

Figure 2

critical temperature, the recrystallisation temperature, i.e. the temperature at which new crystals form at the expense of the old deformed crystals. This recrystallisation temperature is a characteristic of a given material, being roughly proportional to the melting point. It is slightly effected by other factors such as the amount of deformation, type of deformation, and the rate of heating.

Fig. 3 combines the Fig. 1 showing the effect of deformation on properties, with the effect of annealing the deformed material on the properties.

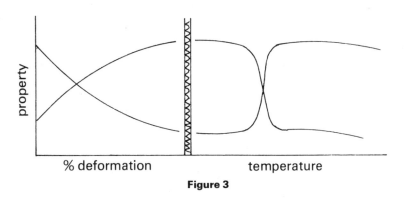

Figure 3

It shows that annealing a deformed material at temperatures up to its recrystallisation temperature has a negligible effect on the properties, but at the recrystallisation temperature there is a very sharp change in the properties.

Starting with the annealed material, deformation leads to an increase in the strength and hardness and a decrease in ductility. Annealing the deformed material below the recrystallisation temperature leads to no significant change in properties, but at the recrystallisation temperature there is a very marked change in properties, the strength, hardness, and ductility returning to their initial annealed values. As the temperature is increased above the recrystallisation temperature there is a small decrease in the strength, hardness and ductility due to grain growth.

This leads to two possible ways of producing a given change in shape. One way is to carry out repeated deformation–annealing cycles, when it is possible to select the appropriate cycle to give the required shape with the required final properties. The other way is to carry out the deformation above the recrystallisation temperature when simultaneous work hardening and recrystallisation will occur; in this case the final shape is produced with the properties of the annealed material. Thus the recrystallisation temperature represents a temperature at which there is a very marked change in the ease of working; if working is carried out below this temperature it is cold working, if above then it is hot working. The distinction can be illustrated by noting that lead hot works at room temperature whilst copper cold works at 200° C.

The choice of hot or cold working is governed by the need to compromise between the requirements of the fabricator and the requirements of the user. Thus ease of fabrication will often favour hot working, but the final properties may require cold working. Frequently the initial deformation is carried out by hot working and the final stages by cold working.

It should be noted that although working is easiest at elevated temperatures, too high a temperature leads to excessive grain growth and a poor surface finish, i.e. the orange peel effect.

Behaviour of Alloys

A great variety of alloys are possible when two or more metals are mixed in various proportions, although only a limited number of these are of any practical value. In most cases the addition of one metal to another initially gives rise to a solid solution; further additions may then exceed the solubility and give rise to a compound, hence giving a duplex alloy consisting of a solid solution and a compound.

a) Solid Solution Alloys

A solid solution is formed by one metal dissolving in another, therefore it is not surprising that it behaves in a very similar way to a pure metal. Thus both copper and cartridge brass, a solid solution of zinc in copper, show the deformation, and deformation and annealing, behaviour of Fig. 3, differing only in the magnitudes of the various properties and in the recrystallisation temperature.

b) Duplex Alloys

One of the most important features of many of these alloys is that the solubility of one metal in the other varies with temperature. Thus in Duralumins, alloys of aluminium and copper, at elevated temperature all the copper is soluble in the aluminium, but on cooling to room temperature the solubility of the copper is reduced causing the copper to precipitate out as the aluminium copper compound $CuAl_2$. Thus at room temperature the equilibrium structure consists of the compound $CuAl_2$ in a solid solution of copper in aluminium. The basis of treatment for this type of alloy is to use various heat treatments to control the precipitation of the compound.

The alloy is heated to the elevated temperature when all the metals are in solid solution, and then it can either be slow cooled to room temperature when the compound precipitates out as relatively few large particles, or rapidly cooled to room temperature by water quenching when the precipitation is suppressed. In the latter treatment one gets the high temperature structure retained at room temperature, whereas the equilibrium-slow-cooling gives the solid solution plus a precipitate. The precipitation is then caused to occur in the quenched alloy either by leaving the material for a time at room temperature or by heating to some temperature below that at which complete solubility exists. When the precipitation occurs by this quenching and tempering (reheating) treatment it does so in the form of a large

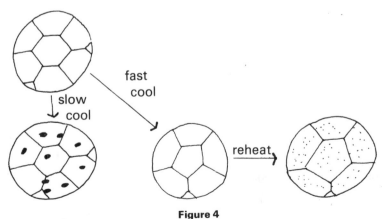

Figure 4

number of fine particles. Fig. 4 summarises these alternative treatments.

Thus by varying the heat treatment of a given alloy, it is possible to vary the morphology (shape, size, and distribution) of a given volume of precipitate, and hence to vary the properties of the alloy.

In the case of Duralumin the initial quenching gives a retained solid solution which can then be deformed to give the required change of shape. Subsequent reheating causes the precipitation of the compound with a considerable rise in strength and hardness but a drop in ductility—this is referred to as age hardening or precipitation hardening.

With the Muntz metal, a copper zinc alloy, a similar treatment is carried out but the improvement in properties is not as marked as in the case of the Duralumin—in this case it is referred to as a quenching and tempering.

Although the quenched alloy can be deformed as described for a solid solution, it cannot be softened by recrystallisation since the required heating is likely to produce precipitation. Thus to soften the hardened alloy it is necessary to heat once again to a temperature when all the precipitate is in solution and then water quench.

Iron and Steels

It will be recalled that pure metals were divided into two groups, one including copper, aluminium, and nickel and the other including iron. The entire range of compositions and heat treatments of steels derive from the existence of a phase change occurring in pure iron at 910° C. Thus the low temperature form of iron, ferrite, has an extremely low solubility for carbon compared to the high temperature form, austenite. For example, a typical medium carbon steel with 0.4 per cent carbon will have all its carbon in solution in the high temperature austenite but only a very small amount in solution in the low-temperature ferrite, therefore on cooling from an elevated temperature the austenite will transform to ferrite with the precipitation of the excess carbon as the compound cementite. This transformation and the associated precipitation can be controlled by heat treatment giving rise to a wide variety of possible properties in any one given alloy.

It would be quite inappropriate in an introduction such as this to attempt to discuss anything but the very basic metallurgy of steels, hence the following refers to a medium carbon steel. The principles involved apply to all steels, but there may be complicating factors which will not be discussed here.

At temperatures above around 850° C. a medium carbon steel consists of austenite with all the carbon in solution. Slow cooling, i.e. annealing, allows the austenite to transform to ferrite with the carbon precipitating out as cementite. Fast cooling, i.e. water quenching, suppresses this precipitation and hence leads to the austenite transferring to martensite; this martensite is a hard and brittle constituent, and its formation is frequently associated with cracking of the component. If this martensite is now tempered, i.e. heated, to some temperature below that at which the austenite forms, i.e. 723° C. in plain carbon steels, it will decompose to give the equilibrium structure of ferrite and cementite.

It can be seen that both slow cooling and quenching and tempering give rise to ferrite and cementite, but the morphology (shape, size and distribution) of the cementite in the ferrite will be very different, and hence the properties will be very different. Generally, the quenched and tempered alloy will be stronger and more ductile than the slow-cooled alloy, whilst increasing the tempering temperature in the quenched and tempered alloy will lead to an increased ductility and a decreased strength.

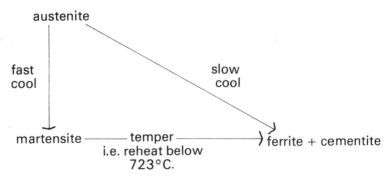

Figure 5

It will be noted that the actual heat treatments are similar to those described for Duralumin and Muntz metal. However, in these alloys the initial quenching suppressed precipitation causing the high-temperature condition to be retained at room temperature, whereas in the 0.4 per cent carbon steel the initial quenching gave rise to a new product, martensite, rather than retaining the high-temperature austenite at room temperature.

The addition of alloy in elements alters the temperatures at which various changes occur, and also the rates of cooling necessary to bring about a given change. For example in stainless steels chromium and nickel are added so making the austenite stable at room temperature, and in high-speed tool steels various additions are made such that the initial quenching to reduce the martensite can be done in oil rather than water, thereby preventing quench cracking.

Welding

Welding involves the localised melting of the material and is likely to be associated with changes in structure and hence properties. For example, if a copper component has been produced by cold working it will be in the work-hardened condition; if now it is welded into some larger component then in the vicinity of the weld recrystallisation will have occurred with a resultant softening. Besides the obvious variation in mechanical properties there will also be an increased tendency to corrosion in the region between the weld and the unaffected material.

These few comments on welding are added to remind you that selecting the appropriate way of joining two components involves not only the choice of the actual joining method but also the effect on the properties of the material as a consequence of the joining method.

*   *   *

This introductory chapter has introduced some of the metallurgical principles involved in the processing of metals and alloys. It is hoped that it will help the 'craftsman' to bridge the gap between his world and that of the professional metallurgist.

1 THE FERROUS METALS

Steel, in its various forms, is essentially an alloy cf iron and carbon, and its qualities may be enhanced, and in some cases new qualities imparted, by the addition of other elements in conjunction with suitable heat treatment processes. By these methods, improved tensile strength, toughness, hardness, uniformity of quality, heat, corrosion and wear resistance can be achieved, and distortion due to heat treatment can be reduced.

Modern-day steels are so complex both in composition and function that it is no longer possible to speak in terms of 'spring steel', 'carbon steel', or 'tool steel' and still convey the precise meaning, for there now exist about twenty different compositions of steel, all of which are suitable for the manufacture of specific types of springs, and many of which will do for one type of spring, and this is true of the carbon steels and the tool steels.

For the same reasons the old and well-known method of identification by the 'spark test' is no longer reliable, although it can still be used to determine cast iron, wrought iron, mild steel, medium and high carbon steels, manganese and stainless steel, and for this reason a spark test identification chart has been included in this pamphlet, although, where possible, alternative methods of identification have also been given.

It is therefore important to be specific in the description of the material under reference, and to this end an indication of the properties that are imparted by some of the elements to the steel of their inclusion, is given below.

2 THE INFLUENCE OF ALLOYING ELEMENTS

Carbon (C): An essential element of alloy steels, it is usually present in quantities of below 0.45 per cent except in high carbon tool steel where it might vary between 0.60 and 1.8 per cent. Usually, the higher the content of the other elements the lower the content of the carbon, but in all cases the ratio of the carbon content to the content of the other constituents is critical.

Chromium (Cr): The addition of this element to steel results in the formation of chromium carbide, and increases the hardening power, and the ability to deep harden, as well as decreasing the tendency to warp in heat treatment. It also increases the resistance to shock and corrosion. Its content varies from 0 to 18 per cent, depending upon the application of the steel.

Cobalt (Co): By the inclusion of cobalt in varying amounts from 0.5 to 14 per cent the effect is to increase the tensile strength, refine the grain of the metal, and to enhance the resistance to heat.

Manganese (Mn): Manganese may be included in amounts varying from 0.4 to 2.0 per cent and 11 to 16 per cent, the lower percentages increasing hardness, resistance to wear, tensile strength, and lowering the melting point, whilst at the higher range the steel is rendered practically non-magnetic, and increases its work hardening properties.

Molybdenum (Mo): The addition of this metal to steel results in increased resistance to wear and to heat, and higher tensile strengths, and renders the steel susceptible to deep hardening. It is included in amounts varying from 0 to 9.0 per cent.

Nickel (Ni): The percentage added is from 0 to 8 per cent, and inclusion results in increased tensile strength, toughness and hardness, and in the high percentage range gives excellent corrosion resistance, and resistance to oxidation at high temperatures, and renders the steel almost non-magnetic.

Silicon (Si): The influence of silicon is to increase the tensile strength and hardness, and to increase the elastic limit, a very desirable property in spring steel. The percentage varies from 1.0 to 2.2 per cent according to the service requirements of the steel.

Tungsten (W): Tungsten, now known as wolfram, might be included in amounts varying from 0.4 to 10 per cent, and its effect is to refine the grain, and to increase heat resistance, wear resistance, shock resistance, and tensile strength. In high-speed tool steels it might be included to the extent of 22.00 per cent.

Vanadium (V): The addition of this metal varies between 0.15 and 2.0 per cent and results in refined grain, increased fatigue resistance and tensile strength, and reduced distortion during heat treatment.

It will be seen that there is a considerable overlapping of the characteristics produced, yet every element has its specific place in making steel to certain specifications, and often two, three or more of the above will be included in one alloy, and their influence on the steel will, as a result of their admixture, impart special qualities. Their ratios to each other are critical in the control of the required specification.

There are even some instances where the inclusion of generally undesirable elements such as sulphur will give a desirable characteristic, and the inclusion of this chemical will give the steel free machining qualities which improve with bright drawing and cold working, although parts made from such material are restricted to low duty.

It is important, too, to realise that the optimum qualities of an alloy steel are only obtained by carefully controlled specific heat treatment, intended to soften, anneal, case harden, deep harden or toughen, and might consist of quenching in air, water or oil at temperatures varying between 780° C. and 1080° C., and tempering at between 550° C. and 780° C., depending upon the composition of the steel. There are, however, one or two steels that require tempering temperatures as low as 200° C. such as 2 per cent nickel–chronium–molybdenum steel, and some that soften on quenching from high temperatures instead of hardening, such as high nickel–chromium–tungsten steel and high chromium–nickel–tungsten steel and high manganese steel.

From this it will be realised that in straightening or bending damaged components made from heat-treated alloy steels, such as steering arms, linkage, axle beams, etc., on motor vehicles and tractors, implement beams, linkage, power take-off shafts, etc., on agricultural machines, this must be done cold, as the temperature gradient through the components, if heated for bending or straightening, will always produce an area that has been subjected to the worst possible conditions of heat treatment which will surely result in failure.

This also holds true for the welding of alloy steels, and it is the failure of sections weakened by adverse heat treatment incidental to welding that is responsible for the saying that the weld is stronger than the original material. It is therefore essential that where high-strength repairs are required, such alloy steels must be fully and appropriately re-heat treated.

3 CAST IRON

Uses

All types of castings not subject to tensile stress, or shock loads such as engine blocks; machine-tool frames; plough-wheel bearings; gear-box housings; mower frames, etc.

Composition

Iron 93 to 96 per cent plus carbon, phosphorus, sulphur, silicon and manganese in variable quantities to make 100 per cent.

Properties

Cast irons can be white or grey, depending on silicon content and rate of cooling. The higher the silicon up to 3 per cent the darker the iron, and the faster the cooling the whiter the iron. White cast iron is very hard and durable although also brittle, and in the form known as chilled cast iron is used for mill rolls and grinding rolls. It is difficult to machine and breaks with a white closely granular structure of hard cementite.

In grey cast iron the colour is derived from the free carbon present as flakes of graphite, and since graphite is soft and a good lubricant, the iron itself is softer, readily machinable, less brittle, and is almost self-lubricating on sliding surfaces.

It is easily identified by the spark test, and by its coarse grey granular structure when fractured, or by its characteristic powdering when drilled, which powder, because of its graphite content, will write as the 'lead' in a pencil.

Welding Procedure

This material can be full fusion welded with cast iron or high silicon rods, or bronze welded with bronze rods. It also welds well by electric arc using basic (low hydrogen) electrodes, and rods with a high nickel content, when, except for rigid and complicated castings, pre-heat should not be necessary. When welding large complex castings by the gas process it is necessary to pre-heat the work to cherry red all over before commencing welding, i.e. 900° C. or 1652° F., after which the whole piece must be cooled evenly and slowly to prevent stressing or cracking.

4 MALLEABLE CAST IRON

Uses

All types of castings subject to stress and shock loads, such as trip, cam and binder mechanisms; brackets; depth adjustment racks; mining machinery, etc.

Composition

As for white cast iron, this being the first stage in the process, the castings then being annealed by one of the two methods described below.

Black Heart Process

Castings heated slowly in an annealing furnace to a temperature of 1000° C. (1832° F.) and held from three to six days and then cooled in the furnace. This results in the cementite breaking down into ferrite and temper carbon which makes the iron more ductile.

White Heart Process

Castings packed in hematite ore in the furnace and heated to 900° C. (1652° F.) which is maintained from five to six days, then cooled slowly in the furnace. This breaks the cementite down into free carbon which is oxydised out by the hematite, so that the casting, now low in carbon, is softer and ductile.

Properties

Malleable cast iron can be bent, twisted and machined, and stands shock loads better than mild steel, and has a tensile elastic limit of 12 to 16 tons per sq. in.

Malleable cast iron can be identified by its surface appearance, which resembles a rough glazing, or by the spark test on the grinding wheel.

Welding Procedure

It must not be fusion welded, but should be bronze welded or brazed. For electric welding use bronze rod with A.C. or D.C. current with ioniser inserted in the circuit, or with basic (low hydrogen) or M/S electrodes using the lowest current possible to maintain the arc. If possible, normalising should follow.

5   WROUGHT IRON

Uses

Ornamental ironwork; railings; chains; railway couplings; hooks, etc.

Composition

Wrought iron, or puddled iron as it is sometimes called, is at least 99 per cent pure iron, and this is achieved by oxidising out the impurities in grey pig iron, i.e. silicon, manganese, sulphur and carbon, by the use of slags and gasses with oxidising characteristics, such as ferric oxide and silica fettled in from burnt pyrites. Four stages of manufacture are involved: 1, the melting stage; 2, the boiling stage; 3, the finishing stage; 4, the balling stage. It is this last stage that gives the metal its characteristic fibrous appearance, for at this stage the metal is now a malleable iron and in a spongy mass, from which is worked balls of 60 to 80 pounds in weight which are removed from the puddling furnace dripping with slag, and forged under the shingling hammer into rough blooms. This forging operation squeezes out a large proportion of the slag, and welds the iron. The temperature of the blooms is then evened up in a soaking pit and then passed through rolls to produce what are called 'Merchant Bars'. Better quality wrought iron bars are made by cutting the Merchant Bars into short lengths, piling them into faggots, re-heating to welding heat, re-forging and re-rolling. Repeated forging and rolling gives successively better steel.

Properties

A fibrous and ropy steel with an ultimate tensile strength of 23 to 24 tons per sq. in., it has a yield point of 15 to 16 tons per sq. in. and a 25 to 35 per cent elongation.

It resists corrosion to a remarkable degree and will withstand repeated shocks at intervals long enough to effect recovery to its normal condition.

Its workability can be severely affected by the inclusion of minute quantities of phosphorus or sulphur, or silicon, resulting in the metal being either 'cold short' or 'hot short', that is being liable to fracture when cold, or to fracture when hot, but possessing the usual qualities at the other range of heat.

It machines badly as a result of its construction and tends to separate along the roaks on all machining operations except in drilling. Otherwise it can be manipulated with ease and precision, and is unsurpassed for work requiring upsetting, splitting, punching, and scarf, cleft and pocket welding. It is the material *par excellence* for decorative wrought ironwork.

Wrought iron can be identified by the spark test, or by twisting and bending a small section and observing the separation of the laminations.

It cannot be hardened by heating and quenching, and is not improved by heat treatment.

Most modern so-called wrought ironwork, although 'wrought' in the sense that it is 'worked', is made from mild steel and not from wrought iron.

Welding Procedure

Wrought iron can be welded by conventional methods and materials as for mild steel, but it is important to weave the weld so as to avoid forming the bead along a roak. It cuts badly by oxy-acetylene torch.

6   MILD STEEL

Uses

Structural steel as in angle, channel and I and T sections; trailer chassis; machine-tool frames; agricultural implement frames; and is probably the most widely used of all engineering materials.

Composition

The term steel is generally applied to those alloys of carbon and iron, where the carbon is present entirely in the combined condition, and is less than 2 per cent in total content, and mild steel will contain never more than 0.25 per cent of carbon, but will be contaminated, as are all commercial quality steels with manganese, sulphur, silicon and phosphorus in minute amounts.

Properties

Soft, ductile, and easily worked either hot or cold. Forges well under heat, and machines easily on all machining operations, but free machining qualities can be obtained in which there is a small amount of lead from 0.15 to 0.30 per cent incorporated.

Mild steel cannot be hardened by merely heating and quenching, and this fact provides a ready means of identification, in that its reaction to filing will be the same after such treatment as it was before. The spark test will also help in identification.

Hardening is effected by the cyanide process or by soaking at red heat long enough for it to absorb additional carbon from such materials as charcoal, leather, or other carbonaceous elements, either solid, liquid, or gaseous, with which it is in contact, which will enable a thin skin of the surface of the metal to respond to heat treatment, the inner core maintaining its normal characteristics. This is known as case hardening.

Welding Procedure

Mild steel can be welded with gas, using copper-coated mild steel rods, S.M.1 or $3\frac{1}{2}$ per cent nickel rods. For electric arc welding excellent results are obtained with mild steel or basic electrodes.

26

M.I.G. welding is now a popular method of joining mild steel using a standard double deoxidised wire and CO_2, Argon-Oxygen or Argon-CO_2 Gas Shield. Because of its qualities it requires no further treatment after welding.

7   MEDIUM CARBON STEELS

Uses

Smith's tools; chisels; setts; dies; shear blades; mason's tools; hammers; miner's and quarry drills; springs; spades; shovels; forks; ploughshares; beams and mould boards; structural steel, etc.

Composition

From 0.25 to 0.60 per cent carbon, and will contain silicon, manganese, sulphur and phosphorus in minute quantities, and iron to 100 per cent.

Properties

Carbon renders the steel susceptible to hardening, although a carbon content of below 0.1 per cent will show no significant hardening when rapidly cooled. An increase in the percentage of carbon results in a proportionate increase in hardness and tenacity, with a corresponding drop in the toughness and ductility.

It can be identified easily by the spark test, and the higher the carbon content the less vivid and smaller are the sparks, but more abundant closer to the grinding wheel.

The maximum forging temperature is 1150° C. and should be finished at 700° C., and the temperature for annealing is 770° C. whilst the hardening temperature for quenching is 800° C. to 820° C. The tempering heat depends on requirements, and the higher this heat the greater is the reduction in hardness with a corresponding increase in the ductility. For this class of steel the tempering heat varies between 220° C. and 300° C.

Welding Procedure

Medium carbon steels can be gas welded with $3\frac{1}{2}$ per cent nickel S.M.1 rods, and electrically welded with basic or austenitic electrodes, but in the case of springs a hard surfacing electrode of 250 Brinell hardness should be used, peening the weld whilst hot to improve its quality and fully hardening and tempering the spring afterwards.

8   HIGH CARBON STEELS

Uses

Special turning and planing tools; twist drills; cutters; mills and hobs; taps; dies; punches; reamers; springs, etc.

Composition

0.60 to 1.8 per cent carbon, but usually limited to a maximum of 1.5 per cent carbon except in special purpose compositions. There is the usual contamination from silicon, manganese, sulphur and phosphorus, the balance to 100 per cent being of iron.

Properties

Susceptible to extreme and deep hardening by the appropriate heat treatment, but must not be used in shock applications. It will consequently grind to a very keen edge that will not be easily dulled.

Identification by the spark test is perhaps easiest of all, the higher the carbon content the greater the profusion of sparks.

It can be forged hot and is quite machinable in the annealed state, but careful treatment is necessary as the grain structure can easily be affected. Maximum forging temperature is 850° C. and forging should finish before falling to 700° C. Hardening is effected by quenching at temperatures between 730° C. and 780° C., and, after grinding off the carburised surface to a bright finish to reveal colour changes, can be tempered at 230° C.

Welding Procedure

High carbon steels can be welded by oxy-acetylene gas using $3\frac{1}{2}$ per cent nickel S.M.1 rods, and by the electric arc process with basic or austenitic electrodes, but should always be pre-heated to about 350° C. before commencing welding. The necessity for welding this class of steel rarely arises, but should it do so, the qualities imparted by careful heat treatment would have to be restored.

9   HIGH-SPEED TOOL STEEL

Uses

Restricted entirely to cutting bits for machine tools.

Composition

Carbon 0.65 to 0.80 per cent, silicon 0.25 to 0.30 per cent, manganese 0.20 per cent, chromium 3.0 to 5.0 per cent, wolfram (tungsten) 14.0 to 22.0 per cent, vanadium 0 to 1.5 per cent, molybdenum 0 to 0.5 per cent, cobalt 0 to 12.0 per cent, with sulphur and phosphorus as low as possible, and iron to make 100 per cent.

Properties

The principal advantage of these steels lies in their ability to retain their cutting edges even when operating at speeds that will produce temperatures of red heat. It is, however, pointless to use them on machine tools that have not the necessary power for driving at high speeds, or the rigidity to prevent vibration from the deep cuts of which these tools are capable.

High-speed steels are readily machined in the annealed state, or forged to shape as lathe, planer, or shaper tools, but will be efficient only if great care is taken in subsequent heat treatment, which demands very high temperatures.

Forging heat is between 1050° C. and 1200° C. and must not be allowed to fall below 950° C., at which temperature re-heating is necessary. Normalising is *essential* after forging, and is done by re-heating to 700° C. for half an hour and then cooling in still air.

For hardening and tempering in the hearth, heat the tool nose slowly, turning frequently until bright red. Increase the blast to raise the temperature quickly until the tip of the nose just begins to run. Withdraw immediately and cool the whole tool in an air blast, or quench in oil. Temper at 230° C. first, and for better results temper again at 580° C.

10 MEDIUM MANGANESE STEEL

Uses

Rifle barrels; H.T. bolts; volute, laminated, coil and torsion bar springs; railway and tram lines; conveyor tracks; structural steelwork; axles.

Composition

Varying very widely from 0.4 to 1.8 per cent manganese alloyed with other elements such as nickel, chromium, carbon molybdenum, vanadium, and silicon, and are often not classified as manganese steels, because manganese might not be responsible for the principal characteristic of the metal, and in any case all modern steels contain some manganese, if only as a de-oxidiser and to combine with the silicon to prevent formation of ferrous silicate, which is always a source of trouble when re-smelting.

Properties

Variable, according to composition and ratio of constituents, but developing from 45 to 65 tons tensile.

They machine reasonably well by conventional methods, and can be forged under proper conditions.

Heat treatment in water or oil or either is dependent upon the composition, but hardening in oil or water takes place from a temperature of 850° C. to 900° C., and tempering from between 550° C. to 720° C.

It can be identified by the spark test, the sparks being very distinctive and emitting a slight hissing sound.

Welding Procedure

If manganese steels are subject to intense heat their structure and characteristics will undergo a severe change, therefore they must be welded at as low a heat as possible, and for this reason they should not be gas welded.

When arc welding such steels the current values should be about 25 per cent less than for equivalent sizes of M/S electrodes, and it is advisable to use the lowest current that will produce a satisfactory weld, allowing time to cool between each run.

Austenitic chrome—nickel alloy electrodes are used with the best results, and maximum current values quoted on the packet label should never be exceeded. The weld metal is soft and ductile, resists cracking on cooling, but can be work hardened by peening if desired.

11 HIGH MANGANESE STEEL

Uses

Soil and rock ripper teeth; digger bucket teeth; dredger bucket teeth; ore and rock crushing jaws and rolls; bulldozer blades; railway points; screenings; lifting magnet cover plates; safes and vaults; soldiers' helmets, etc.

Composition

1.0 to 1.4 per cent carbon, 1.0 to 1.8 per cent silicon, 10 to 16 per cent manganese, with minute quantities of sulphur and phosphorus.

Properties

Discovered by Robert Hadfield in 1842 it is remarkable for its extreme toughness and the fact that its surface hardness increases with repeated impact.

This steel is non-magnetic, which quality provides a ready means of identification.

It can be hot forged with ease, but is very difficult to machine and even in its most ductile form requires special tipped tools.

Maximum ductility and toughness are obtained by heating to 1000° C. (1832° F.) and quenching in water, the process being the opposite to that used for carbon steels.

Although in the past it has not been widely used in agriculture, today, with increasing emphasis on mechanisation, it is gradually finding its place, due to its remarkable resistance to abrasion, and for that quality alone it is likely to be used more and more on the farm.

Welding Procedure

It cuts badly with the oxy-acetylene torch, and is generally unsatisfactory to weld by the same means, but it can be readily welded by electric arc using austenitic electrodes with a 14 per cent manganese content, or 18/8 stainless steel electrodes.

12   STAINLESS STEELS

Uses

Kettles; autoclaves; hospital and domestic equipment; chemical and food processing and storage plant; milking apparatus; bearing races; turbine blades; pumps shafts, etc.

Composition

18 per cent chromium and 8 per cent nickel known as 18/8 stainless, and 12 per cent chromium and 12 per cent nickel known as 12/12 stainless, and 18 per cent chromium and 9 per cent nickel known as staybrite. These are known as the chromium nickel steels, but there are many other varieties classified as plain chromium steels and others as high chromium low nickel steels which will not be considered here.

Properties

The 18/8, 12/12, and 18/9 group cannot be hardened by heat treatment and are non-magnetic. They often develop a tensile strength of up to 90 tons per sq. in., and may contain small amounts of tungsten, titanium, molybdenum, copper, carbon, etc. Cold working will increase the tensile strength, and hardness.

They can be worked cold, pressed, formed, bent, welded, brazed, and soldered, and free machining qualities can be imparted by the addition of selenium or by a high sulphur content.

It is not possible to mistake this steel for anything other than in the stainless range, but when applied to a grinding wheel the sparks are long, bright yellow streaks terminating in spear points; also, in their soft state they are completely austenitic and non-magnetic, and are softened by quenching from heats at 1100° C. (this process being opposite to that used for carbon steel) and this may have to be done several times during cold working, and on completion of manipulation in order to restore softness, ductility and maximum resistance to corrosion, which latter is only achieved in the softest state.

Welding Procedure

Satisfactory welding by gas can be achieved by using stainless steel rods, or bronze welding with 9 per cent nickel bronze rods. Columbium stabilised stainless steel electrodes produce good results in electric arc welding.

Production welding of stainless steels is normally carried out using T.I.G. welding on thinner sections where weld appearance is important and M.I.G. welding on thicker sections where production speed is important.

In both methods choice of filler wire is important. The joint to be welded must be free from grease and scale and removal of any scale produced by the welding process is also important.

Argon is used as the shielding gas to protect the arc; a backing gas will normally have to be applied to the back of the joint to prevent oxidisation of the underside of the weld.

13   SILICO-MANGANESE STEELS

Uses

This is essentially spring steel, but has some uses in the field of gear transmission.

Composition

0.45 to 0.55 per cent carbon, 0.60 to 0.90 per cent manganese and 1.80 to 2.20 per cent silicon, sulphur 0.045 per cent max., phosphorus 0.045 per cent max. and iron to 100 per cent.

Properties

The best springs are made from materials that can store up the greatest amount of energy in a given section of the material, without permanent deformation or failure; they should also possess maximum resistance to fatigue and shock effects, and have a correspondingly high deformation or deflection and recovery value. For these qualities it is essential the material should have as high an elastic limit as possible. The silicon content is responsible for this high elasticity.

Forging temperatures are 1050° C. to 1100° C. down to 900° C., and it is heat treated by quenching in oil at its minimum forging temperature (900° C.) and tempered at 500° C. to 550° C.

Welding Procedure

Welding by electric arc is satisfactory but bevelling to a double V and pre-heating to 300° C. is essential, and a ferritic-type rod should be used, the entire spring being re-hardened and tempered on completion to eliminate the brittle zone adjacent to the weld. If, however, this is impossible, local tempering of the weld zone can be accomplished by very careful heating to 3 or 4 in. either side of the weld to 500° C. to 600° C. by the use of a blowpipe, using chromatic thermometers (coloured crayon heat indicators) for reliable guidance of the applied heat.

14   NITRALOY

Uses

For extremely severe service, particularly suited to gearing and where great resistance to wear is important, such as on sliding surfaces as in the cylinders of I.C. and C.I. engines, recoil mechanisms of guns, track pins on crawler tractors, etc.

Composition

Classes as a chromium–aluminium–molybdenum steel, a typical analysis would be carbon 0.36 per cent, manganese 0.51 per cent, silicon 0.27 per cent, aluminium 1.25 per cent, chromium 1.49 per cent, molybdenum 0.18 per cent with the minimum sulphur and phosphorus contamination, and iron to 100 per cent.

Properties

Specially formulated on the basis of the fact that certain steels will absorb nitrogen, which enables extremely hard non-brittle surfaces to be produced, the nitriding process is carried out in an electric furnace at 510° C. which temperature is critically controlled, and the parts are placed in a gas-tight box with inlet and outlet tubes for circulation of ammonia gas. The process takes from 2 to 90 hours depending on the depth of hardness required. Parts or areas required to be kept soft, can be, by shrouding them in tin, nickel, or copper plating.

Nitraloy steels may readily be machined in the heat-treated as well as in the annealed state, but they are virtually unmanageable prior to stabilising after forging, and must be thoroughly relieved by annealing at 537° C. before attempting to machine, or annealed and heat treated. This annealing must also be done prior to nitriding. Close tolerances can be maintained in finish machining, as due to the low heat of nitriding very little distortion occurs.

The usual sequence is anneal, rough machine, heat treat, final machine, nitriding, grind to finish. *Or*, if forged to close limits, anneal, heat treat, machine, nitriding, grind to finish. High-speed steel or tipped tools are necessary for machining. Nitraloy will not gas weld, and electric arc welding destroys its properties, which can only be re-established by a properly equipped heat-treatment plant.

15   A BRIEF CONSIDERATION OF OTHER STEELS

The foregoing is by no means representative of the steels available to industry today, but is merely a guide to those steels that a rural workshop will most likely have experience of at some time or other, and has been designed to help the craftsman in the identification and manipulation of them.

It would be imprudent, and misleading, to make no mention of the very widely diversified varieties of other alloy steels, some used as substitutes for those already mentioned, and some superior to them in every way possible for specific applications, the limiting factors resting with the cost of production, or the difficulties of processing or machining.

A typical example is chrome–vanadium steel, limited to the manufacture of high-grade laminated, coil and volute springs and auto and aircraft engine valve springs, it has many characteristics identical to silico-manganese steel but exceeds the elastic limit of the latter by 30 per cent and the Izod impact rating by 80 per cent.

Another spring steel is silico-chrome steel, not as good as silico-manganese, but far superior to high carbon spring steel.

Nickel–chromium–molybdenum steels are used for gears, but, as previously stated, silico-manganese steel is also used for this purpose, while silico–chrome and nickel–chrome and even high-speed steels are used for auto and aircraft engine valves.

A great variety of low-carbon steels exist, with which are alloyed one or several of the alloying elements in varying quantities, and consequently of widely varying characteristics, and different designers of equipment will make slightly different uses of these materials. Almost all of them, with or without the addition of up to 1.10 per cent aluminium, can have their normal characteristics completely changed but improved by subjecting them to the nitriding process, so making them suitable for very severe duty on components such as high-speed crankshafts, airscrew shafts, engine valves, and endowing them with qualities of hardness retention up to 500° C., whereas ordinary case-hardened steels begin to lose hardness at temperatures as low as 120° C.

This process of ever-developing new types of steel and new methods of improving their quality has been going on to our knowledge for at least 2900 years, for Homer (probably 900 B.C.) writes in his *Odyssey*: 'And when a Smith dips an axe or adze in chill water with a great hissing, then he would temper it, for whereby anon comes the strength of iron.'

But it is only in the last 100 years that man has come to understand 'how' and 'why'.

Further reading:

Engineering Materials. Vol. 1: A. W. Judge. Sir Isaac Pitman & Son Ltd.

Hints on Steel. Arthur Balfour & Co. Ltd.

Welding Manual. The Quasi-Arc Co. Ltd.

16 THE NON-FERROUS METALS

These metals, grouped under the above heading because they contain virtually no iron, cover a very wide range, including the precious metals such as gold and platinum, but only those in more common use in industry will be considered here.

As with the steels, the non-ferrous metals are imparted special properties by alloying them, often with elements common to the alloy steels, such as manganese, nickel, and silicon, etc.; and often by alloying two or more non-ferrous metals together in varying proportions.

It is often a source of surprise to learn that the inclusion of a weak metal may produce an alloy much stronger than the strongest of the metals in the alloy, and indeed this is true of the steels, which by the inclusion of weak and brittle carbon in iron will raise the Ultimate Tensile Strength from 8 tons/sq. in. to 27 tons/sq. in. and more.

Similar striking examples are the inclusion of aluminium in copper and copper in aluminium, and it is noteworthy that 6 per cent of aluminium in a chill cast copper–aluminium alloy raises the Tensile Strength of the copper from 12 tons/sq. in. to 18 tons/sq. in.; 10 per cent aluminium raises the strength to 36 tons/sq. in., but after that any increase in the percentage of aluminium causes a sharp fall in the Tensile Strength, whilst aluminium alloyed with up to 12 per cent copper shows a considerable if not as spectacular an increase in strength.

Another peculiar phenomenon lies in the fact that when one metal is alloyed with another the melting point is always affected, but not necessarily in accordance with the ratio of constituents. In some cases the melting point of an alloy can be arrived at by simple arithmetic as with copper nickel alloys, but in other alloy systems only experiment will reveal the melting point, which might be very much lower than the metal with the lowest melting point temperature in the system.

As an example, lead melts at $327°$ C. and tin at $232°$ C. but if these two metals are alloyed in the ratio of 62 per cent tin and 38 per cent lead the melting point will be $183°$ C.

From this it will be seen that by very careful choice of constituents it would appear possible to make alloys of useful strength and very low melting points. Such alloys are known as Eutectic Alloys or Mixtures, the word eutectic being derived from the Greek *eutektikos*, meaning 'easily melted'.

Because non-ferrous metals can be alloyed with each other in widely differing proportions for widely differing duties, it is necessary to treat each as a separate subject under the heading of the parent metal.

17   ALUMINIUM

Aluminium is derived from bauxite, a bright red earth found in Jamaica, British Guyana, America, West Africa, etc., in rich enough alumina content and economic quantities for processing, although about 8 per cent of the earth's crust is formed of aluminium silicates. One hundred years ago aluminium was considered more precious than gold, and cost £7 per ounce, and cutlery and plate made from it were used at the tables of European kings, and Napoleon III allowed it to be used only on special occasions.

Before 1883, because of its very high smelting temperatures, it could not be processed commercially into aluminium, but in that year, with the introduction of the electrolitic process and the use of cryolite as a flux, the commercial production of this metal became possible.

The aluminium alloys can be divided into two main groups, the wrought and the cast, the first consisting of sheet, strip, wire, and extrusions, and all those other forms attained by 'working', whilst the second consists of alloys in the 'as cast' form such as crank cases, cylinder heads, various brackets, mounts, impellers, etc., and the die cast form.

It is of interest that, although the forging or stamping of aluminium follows the same technique as for steel, the drop of a hammer must be 40 per cent heavier for this softer and lighter metal in order to produce a sound consolidated product.

It is important to note that copper alloyed with aluminium is referred to as copper–aluminium alloy, whilst aluminium alloyed with copper is called aluminium bronze.

As an example of its resistance to corrosion, Eros, the famous statue in Piccadilly Circus, after nearly seventy years of exposure to London's highly corrosive air and the vandalism of high-spirited revellers, is still almost as good as new.

18   ELEMENTS FOR ALLOYING WITH ALUMINIUM

Chromium (Cr): Gives increased hardness and strength. Improves corrosion resistance.

Copper (Cu): Decreases shrinkage and hot shortness and provides the basis for age hardening, but other elements must be present, such as silicon and magnesium, for response to heat treatment.

Iron (Fe): Is always present in aluminium and acts as a hardener, but inhibits age hardening in aluminium copper alloys unless sufficient silicon is included.

Magnesium (Mg): When added to molten metal removes oxides and other impurities, and can be used to neutralise the action of the iron on the copper, and with silicon contributes to hardness.

Manganese (Mn): The inclusion of manganese in aluminium results in a dense structure and even grain, and increases strength.

Nickel (Ni): Gives an alloy with good high-temperature properties.

Silicon (Si): Is always present in aluminium and should be in sufficient amount to nullify the effects of iron, and with magnesium contributes to hardness.

Titanium (Ti): Very high melting point (1800° C.) metal which toughens aluminium alloys and refines the grain.

Zinc (Zn): Is added in amounts up to 10 per cent to improve the mechanical properties.

19   COMMERCIAL ALUMINIUM, WROUGHT AND CAST

Uses

Lightweight structural sections; optical mirror reflectors; cooking utensils; food containers; dairy processing plant, including piping; weather-proofing; toothpaste, ointment, and hairdressing dispenser tubes; low-strength castings for domestic appliances; engine mountings; gear housings, etc.

Composition

99.2 per cent aluminium, 0.5 per cent iron, 0.3 per cent silicon, although higher purity aluminiums are available for special purposes as conductors, etc.

Properties

A highly malleable and ductile material, and one of the lightest of metals, it can be cast with ease to intricate shapes, or cold worked by spinning, deep drawing, stamping, bending, machining, and impact extrusion. As cast its maximum tensile strength is about 7 tons/sq. in. with an elongation of only 3 to 4 per cent, but as rolled and annealed, although the tensile strength remains approximately the same, the elongation percentage is increased to 30, due to refining of the grain structure.

It will not respond in any way to heat treatment except for softening, which is achieved by heating and quenching, but its strength is somewhat improved by cold working.

By the introduction of small quantities of antimony, lead, cadmium, and bismuth, free-cutting qualities have been evolved that can be machined at very high speeds, as the swarf breaks into small chips, rather than coiling and wrapping around the work and tool, and for automatic, capstan, and turret work production this quality of aluminium is essential.

Welding Procedure

Oxygen and Acetylene Gas: For small repairs, conventional gas equipment can be used. The weld area must be thoroughly prepared and welding commence immediately after cleaning, and a good aluminium flux used on the filler rod or the parent metal. A rod containing from 5 to 10 per cent of silicon may sometimes be an advantage.

A soft flame is necessary due to the low melting point of 650° C., and extreme care must be taken to prevent over-heating and collapse, as no colour change takes place to indicate the temperature; and because it oxidises very readily even without heat under no circumstances must the flame be oxidising, but only the faintest excess acetylene haze should be discernible. Gauge pressures should be somewhat lower than normal for the nozzle size in use.

Castings must be vee'd out to 90° and if the fracture is right through it must be vee'd out half-way through, a bead laid in the bottom, and the vee on the reverse side cut until the new weld metal is reached. The whole job is then pre-heated, when welding can proceed using either an aluminium flux or a puddling rod. All flux must be removed immediately after cooling.

Electric Arc: Although welding can be achieved by this means, a D.C. set must be used with electrode negative polarity and a very high degree of operator skill is necessary.

Shielded Metallic Arc: The most convenient method of welding aluminium.

Preparation is all important. After degreasing oxides must be removed immediately prior to welding, using rotary wire brush or coarse steel wool on thinner sections.

Thinner sections are usually T.I.G. welded, whereas M.I.G. welding may be used on thicker sections and for positional welding.

Pre-heating to 200° C is advisable on sections over $\frac{1}{4}$ in. when T.I.G. welding and over $\frac{3}{4}$ in. with the M.I.G. process.

When T.I.G. welding an A.C. power source is used with a zirconiated tungsten with a hemispherical end.

With T.I.G. and M.I.G. process selection of filler wire is most important and to a lesser extent the gas shield.

20 DURALUMIN

Uses

Due to its very high strength-to-weight ratio this alloy is used primarily for light structural work in air frames and cycle frames, but occasionally it appears as a casting.

Composition

Variable within quite a wide range, but a typical one is copper 3.5 per cent, manganese 0.4 per cent, magnesium 0.4 per cent, silicon 0.4 per cent, iron 0.4 per cent, and aluminium 95.3 per cent.

Properties

The mechanical properties of this alloy as normally supplied by the makers are: Yield Point 16 to 18 tons/sq. in.; Ultimate Tensile Strength 25 to 28 tons/sq. in.; Elongation 15 to 20 per cent; Brinell Hardness 100, and since it specific gravity is only 2.8 compared with 7.86 of steel, this makes it, weight for weight, almost three times as strong.

Duralumin can be rolled, drawn, spun and forged, and is supplied in sheet, tube, bar, strip, wire and extruded form, as well as forgings and stampings. Precise control is essential in heat treatment, but the maximum mechanical properties become evident only over a period of days. This age-hardening process can be accelerated by boiling in water.

Welding Procedure

As for commercial aluminium, using duralumin rods, but heat treatment is essential to restore the metal's normal properties. The procedure is to normalise at 490° C. and this temperature must be precisely controlled. Allow to cool in draught-free air and age at room temperature for at least three days.

21 HIDUMINIUM (RR ALLOYS)

Uses

Developed by Rolls-Royce in the first instance for high-duty aero-engine pistons because of its strength retention at elevated temperatures, there are now many variations of this alloy, but the four in most popular use cover compositions suitable for sand- and die-castings for gear-boxes and differential and back-axle casings (RR50); for die-cast pistons, and with 2 per cent silicon for very delicate die-castings (RR53); for high-duty connection rods and general-purpose forgings (RR56); and for forging special heavy-duty large diesel-engine pistons (RR59).

Composition

	Cu	Ni	Mg	Fe	Si	Ti
RR50	1.30	1.3	0.1	1.0	2.2	0.18
RR53	2.25	1.3	1.6	1.4	1.25	0.10
RR56	2.00	1.3	0.8	1.4	0.7	0.10
RR59	2.25	1.3	1.6	1.4	0.5	0.10

with the remainder aluminium.

Properties

Ideal for the purposes for which developed but limited to these applications, although more and more castings such as cylinder blocks, cylinder heads, gear-boxes and housings in these or similar alloys are being produced in modern motor-car manufacture, and will be increasingly met with in future. A fine-grained, high-strength alloy with strengths above that of mild steel, the hiduminiums are classed as aluminium magnesium alloys and for casting purposes might contain as much as 10 per cent magnesium.

Welding Procedure

As for commercial aluminium, using 5 per cent silicon aluminium rods and aluminium flux.

22 NOTES ON SOME OTHER ALUMINIUM ALLOYS

Many other compositions of aluminium alloys will be met with from time to time, but it is not possible to list and specify them all, and in fact new alloys for new duties are being devised continuously.

No mention has been made of the 'Y' alloys developed by the National Physical Laboratories as far back as 1914. These were discovered to have the property of increasing their strength by as much as 50 per cent by age hardening either in the wrought or cast condition, whereas duralumin castings respond only very slightly.

Clad aluminium sheets are extensively used in the aircraft industry, because duralumin sheet corrodes very quickly in unfavourable atmospheres, and is therefore, in spite of its strength, not the most desirable material for aircraft wing and fuselage covering. To overcome this disadvantage, a process was devised for sandwiching a sheet of duralumin between two protective sheets of pure aluminium, which are highly resistant to corrosion. Such sheets are known generally as 'Alclad'.

23 MAGNESIUM ALLOYS

Uses

Mainly in the aircraft industry due to very high strength-to-weight ratio, their weight being only two-thirds that of aluminium and one-quarter that of steel, but some motor-vehicle manufacturers are finding uses for this material in lightly stressed parts such as gear-box covers, rocker gear and timing-case covers.

Composition

Varies widely according to its duty and may contain from 3.0 to 9.0 per cent aluminium, 0.15 to 2.0 per cent manganese, 0.3 to 1.5 per cent zinc, with lesser quantities of silicon, copper, nickel, iron, calcium, tin, with normally not less than 85 per cent magnesium.

Properties

Magnesium itself has very little mechanical strength and in its pure form, although the third most abundant metal on earth, has little use except in photography, signalling and incendiary devices, but when alloyed with other metals a wide range of properties can be achieved.

As cast, the tensile strength of such alloys might be between 5 and 10 tons/sq. in. but this can be improved to 16 tons/sq. in. in sheet and strip form, and to as high as 22 tons/sq. in. when forged or extruded. Heat treatment can improve resistance to shock.

Strength can be added by careful design of radii, bosses, flanges, and fillets by allowing greater thicknesses at such critical points on components, but in such cases this extra weight rather detracts from any advantage it might have over the aluminium alloys.

Magnesium alloys can be wrought by all ordinary means, but the heat generated in turning must be kept below 600° C. as at above that temperature it is likely to ignite chips, swarf, and dust, which burn very fiercely. Should this occur water must *not* be used to extinguish the fire, and any machine shop engaged in machining this material should have a handy supply of asbestos meal or dry sand for extinguishing ignition.

Magnesium can be easily identified by its inflammability, as a few shavings will flash when touched with the flame of a torch, and will not leave black specks of oxide as aluminium does. They are also usually yellow golden in colour due to chromate treatment.

Welding Procedure

By oxy-acetylene is virtually as for aluminium, using a rod of the same material as the parent metal, different rods being available for sheet and castings. A special flux for magnesium welding must be used and pre-heat to 200° C. is essential for castings, as the coefficient of expansion is very high. Welding will be faster than for aluminium due to the lower melting point of 635° C. and great care must be taken to prevent collapse or sag. Wet asbestos meal packed inside of a casting will prevent this. Flux must be washed off completely as any residue will cause the metal to decay. Castings should always be chromated after welding, and several solutions are available on the market. There is little fear of ignition during welding, unless excessive heat is applied and maintained, which is more likely on finely divided particles than in sheet and castings.

Most of the magnesium alloys may be welded with the T.I.G. or M.I.G. method in a similar manner to aluminium using a compatible filler rod and Argon Gas.

24 COPPER AND ITS ALLOYS

Copper has been used for at least 5000 years for utensils, and decorative and engineering purposes in that part of the world now known as Iraq, but it is known that it was used extensively in Britain long before the arrival of the Romans.

About 100 years ago the major part of the world's requirements of copper was smelted in Swansea, the ores being mined in Cornwall or Spain.

The mad rush in search of gold in America, Chile, and Australia revealed, to the disappointment of many but to the joy of some, large deposits of copper, and the more economic working of these rich belts reduced Swansea's output almost to nil. Further discoveries in Rhodesia, Russia, and Japan made the Cornish workings virtually useless, at least economically.

By far the greatest single use of copper today is for the transmission of electrical power, but its uses vary over such a wide field as to interest electrical, mechanical, and chemical engineers, architects, builders, plumbers, mariners, interior decorators, electro-platers, and a host of others.

It is produced in many grades for many purposes and the most important of these are listed below:

Cathode Copper: This is the direct product of electrolytic refining and is of high purity, being the raw material for the production of electrolytic high-conductivity tough pitch, and oxygen-free coppers, for the electrical industries, and is also used for high-conductivity castings.

Electrolytic Tough Pitch High-Conductivity Copper: Prepared by re-melting cathode copper and casting into shapes required for working into wire, strip, bar, tube, extrusions, and sheet, for use in applications where high electrical or thermal conductivity is important, as in electrical apparatus and heat interchangers.

Fire-Refined Tough Pitch High-Conductivity Copper: Of virtually the same quality as the electrolytic grade, but fire refining is only used where it is not important to recover any precious metals entrapped.

Ordinary Tough Pitch Copper: Three British Standard grades are available. The conductivity is not specified and this copper is used for applications where high electrical conductivity is not required.

Oxygen-Free High-Conductivity Copper: Only produced on a commercial scale in America and Finland by re-melting and pouring cathode copper in an atmosphere of carbon monoxide and nitrogen so that no oxygen can be absorbed, but very low oxygen content copper is made in this country by other means for special purposes. The absence of oxygen makes the copper immune to 'gassing', which makes it suitable for flame welding and brazing and for the preparation of glass-to-metal seals. It is preferred for thin wall radiators and other work involving a heavy degree of cold working, such as very deep drawing or spinning.

Phosphorus De-Oxidised Copper: When phosphorus is used as a de-oxidising agent in copper, oxygen is absent, and the included residue of phosphorus makes the metal specially suitable for welding and brazing and is easy to cast and extrude. It is also widely used in the form of plate, and fabrications requiring welding.

De-oxidised copper with an increased phosphorus content is used as a filler rod for brazing non-ferrous alloys.

Arsenical Copper: Either tough pitch or de-oxidised copper may contain between 0.30 and 0.50 per cent of arsenic in solid solution; this reduces the conductivity but increases the strength and reduces atmospheric corrosion. It is widely used in making fire-boxes, piping systems and chemical plant operating at elevated temperatures, and for domestic plumbing.

Welding Procedure

Oxygen and Acetylene welding can be employed for jointing copper, but because of the high thermal conductivity, pre-heat is essential, and more than one torch might have to be used on large fabrications and castings to maintain the welding temperature. Thorough cleaning and flux is necessary, and a slightly oxidising flame is often an advantage.

Welding rods should be of de-oxidised copper with an increased phosphorus content, but brazing with conventional brass rods and fluxes is satisfactory where high strength and colour matching is not important.

Electric Arc welding is best achieved by D.C. reverse polarity, but electrodes are available for A.C. plants. Because of 1, higher thermal conductivity; 2, higher electrical conductivity; 3, higher co-efficient of thermal expansion; and 4, the lower melting point and greater fluidity, it is necessary to allow for greater root openings, larger groove angles, more frequent tacks, higher pre-heats and higher currents, and this holds true for all copper alloys including the brasses and the bronzes. Copper can also be welded by the T.I.G. and M.I.G. methods. T.I.G. usually predominates, using D.C. electrode negative for all alloys except those containing more than 0.5 per cent AL when A.C. is used. Argon is used normally as the shielding gas. It is not recommended that lead alloys are welded.

Copper is used for many other less-well-known purposes such as gilding, sintering, photogravure, and material printing, and as a 'strike' for silver, cadmium and chromium plating, and for friction surfaces such as clutch- and brake-linings. Gramophone records are produced from copper master discs because of copper's ability to reproduce extremely fine detail, and paint and ink makers are large users. It is also extensively used in the manufacture of artificial silks and the production of copper sulphate so widely used in agriculture.

Once it was extensively used in the kitchen, and one of the finest displays of copper kitchen utensils is on permanent view at the Pavilion in Brighton.

Copper is the 'parent' metal of the brasses and bronzes, and the following list indicates the effect of alloying copper with the elements mentioned:

25 ELEMENTS FOR ALLOYING WITH COPPER

Aluminium (Al): Copper cannot ordinarily absorb in its solid state more than 9.4 per cent aluminium, and more common contents vary between 4 and 7 per cent, although as much as 14 per cent is included in some duplex commercial alloys. Aluminium has the outstanding characteristic of protecting itself from corrosion by forming a very thin but very tough film of oxide even in the liquid state, and it extends this protection to the copper at quite elevated temperatures, the mechanical properties remaining satisfactory.

Antimony (Sb—stibium): Imparts superior wearing qualities to bearing bronzes in contents of up to 1.0 per cent.

Arsenic (As): 0.30 to 0.50 per cent content in copper increases the tensile strength which is maintained at temperatures up to 300° C.

Beryllium (Be): Added to copper in contents ranging up to 1.8 per cent, this element imparts great strength to the alloy. Up to 100 tons/sq. in. tensile, with hardness up to 400 D.P.H. can be attained, but elongation in this condition is only about 4 per cent on $4\sqrt{\text{area}}$.

Cadmium (Cd): In contents of from 0.6 to 1.0 per cent, cadmium strengthens and toughens the metal and increases its resistance to fatigue.

Chromium (Cr): Up to 0.5 per cent of chromium, with suitable heat treatment, markedly improves the mechanical properties, and as the metal depends on heat treatment, and not cold working for these improved qualities, it can be used at moderately elevated temperatures up to 450° C. without risk of softening.

Lead (Pb): Improves the machinability and the content may vary between 0.5 and 3.5 per cent. It has no marked effect on the tensile strength but does impair ductility and impact values. 30.0 per cent of lead is used for some special applications.

Manganese (Mn): The inclusion of manganese in a copper alloy enables the metal to be both work hardened by rolling, drawing, and hammering, and by heat treatment, to give strengths as high as 85 tons/sq. in. tensile, and the copper–manganese–nickel alloys are second best, only by a small margin, to beryllium copper. Manganese is also used as a de-oxidiser, as it is in steel making.

Nickel (Ni): Added to copper in proportions from 10 to 30 per cent results in improved strength and durability and lightens the colour. It is now considered an important constituent of aluminium bronze alloys.

Phosphorus (Ph): Used as a de-oxidiser in the straight copper–tin alloys, it cleanses the metal of gases and oxides, and improves the quality, and if in balance no trace is left in the alloy; but if included in such quantities as to both cleanse the metal and leave a residue, this produces phosphor bronze, which has increased resistance to corrosion, and higher strength and hardness.

Silicon (Si): This is used both as a de-oxidiser and as a major alloying element to improve strength, workability, and ductility, and copper–silicon alloys have exceptionally good hot forging characteristics, and promote weldability, although near the melting range the metal is hot short.

Tin (Sn): The alloying of tin with copper produces bronze, the proportion of tin varying between 5.0 and 16 per cent, although in the higher range the metal must be specially treated to absorb the excess tin into solid solution.

Zinc (Zn): Alloys of copper and zinc form the brasses. Up to 50 per cent of zinc might be included for some industrial purposes. Three main conditions exist in straight copper–zinc alloys. These are the alpha brasses with up to 37 per cent zinc, the alpha and beta phase starting at about 37.5 per cent, and extending to 46 per cent at which the beta phase commences. At over 50 per cent zinc there is in fact a fourth phase initiated, known as the gamma, but this lies outside the normal range of brasses.

Again, as in those elements used in alloying with steel, it will be seen that there is considerable overlapping of the characteristics, and more than one of the above will almost invariably be used in varying proportions in producing a specific alloy.

There are, however, many other alloying elements such as titanium, tellurium, tungsten, silver, cobalt, iron, vanadium, and selenium, and although all play their part, particularly in conjunction with those major alloying elements mentioned above, their roles are minor ones, and are only mentioned in passing.

The true bronzes are alloys of copper and tin, just as the true brasses are alloys of copper and zinc, but due to additional benefits conferred by including zinc in bronze or tin in brass, often with additional elements, this has become standard practice.

26 ALUMINIUM BRONZE

Uses

Very widely used in marine, chemical, and hydraulic engineering; gears; turbine runners; stern tubes; chains; piston heads; anchors; cylinder heads for aircraft engines; pumps; shackles; turnbuckles; swivels; screw down nuts; selector forks; electrical contacts, and non-spark tools.

Composition

Although not a true bronze because no tin is included there are numerous compositions of this alloy, but the two main ones may be classed as the one containing 4 to 7 per cent aluminium and termed a binary alloy, and used chiefly for conversion into wrought forms, and the one containing 7 to 11 per cent aluminium containing also nickel, iron, and manganese in varying proportions, a complex alloy, used mainly for cast parts, and responding well to heat treatment.

Properties

The binary alloys have good ductility, and can be fabricated or formed by both hot and cold working processes. In the as-cast form, a strength of 20 tons/sq. in. can be expected, but this increases proportionate to the aluminium content up to 36 tons/sq. in. With 7 per cent aluminium content this can be increased to over 40 tons/sq. in. by cold rolling, but with an inevitable proportionate loss of elasticity and ductility.

Castings in the 7 to 11 per cent aluminium content range, fully heat treated, will have a strength of 40 tons/sq. in. tensile, a higher yield than manganese bronzes, and the wear and abrasion-resisting properties are much superior to both manganese bronzes and steels.

All of these properties are retained at elevated temperatures, including low creep characteristics and resistance to oxidation. Heat treatment can influence both the structure and mechanical properties very widely, but it must be precisely controlled and specific to the alloy.

Welding Procedure

Oxygen and Acetylene: Clean joint by grinding or scraping, finishing by emery cloth; wire brushing is not enough. Leftward weld, scraping the rod to clean the weld and remove trapped gas. The flame should be quite neutral and never oxidising, because of aluminium's affinity for oxygen. Sheets should not be clamped as long welds must be allowed to shrink, which would cause cracking if rigidly held. Start welding half-way along a seam and weld to the end, going back to the middle and welding the other half in the opposite direction. Rods should always be of slightly greater diameter than the sheet thickness. Keep a large area well heated, and add the filler quickly without deep fusion. A calcium–fluoride flux is essential, and rods should be the same as the parent metal but nickel-bronze may be used.

Arc Welding: Cleaning must be effected by grinding or scraping. For sheets up to $\frac{3}{8}$ in. thick butt jointing is recommended. For plates $\frac{3}{8}$ to $\frac{5}{8}$ in. a V-joint chamfered at 50° for four-fifths of the plate thickness should be used, and for plates over $\frac{5}{8}$ in. thick a double-V joint of 50° each side for two-fifths of the plate thickness, will be necessary.

Except for thin sheets, pre-heating is usually necessary, but often this can be done by holding a long arc for a few seconds so as to warm up the base material at the starting point, but for the higher aluminium grades, pre-heat temperatures should be 360° C. to 430° C. Although electrodes are available for A.C. welding, direct current should be used with the electrode positive, which should be held at 70 to 80° to the seam and given a short side-to-side motion as it is moved along. The weld must be made much faster than with steel as the electrodes melt very quickly.

Interrupted welds should be re-started on the summit of the original bead to regenerate the necessary pre-heat. For material up to $\frac{1}{4}$ in. thick the electrodes should be as thick as the material. Over $\frac{1}{4}$ in. thickness of material, $\frac{1}{4}$ in. or $\frac{1}{16}$ in. electrodes may be used.

Grades with high aluminium contents should be annealed after welding at a temperature of 620° C. followed by rapid cooling in air.

Aluminium bronzes may be T.I.G. and M.I.G. welded successfully using compatible filler wire which should be carefully selected.

27 BERYLLIUM COPPER

Uses

Instrument springs, corrugated diaphragms, flexible bellows, Bourdon tubes, non-sparking hand-tools, plastic moulds, etc.

Composition

From 1.6 to 2.4 per cent beryllium with copper to 100 per cent, but occasionally up to 0.4 per cent nickel or 0.25 per cent cobalt might be added for reasons of economy or to inhibit grain growth.

Properties

The copper beryllium alloys have outstanding hardness and strength with exceptional resistance to wear and fatigue. In the cold rolled and fully heat treated condition, tensile strengths of between 75 and 100 tons/sq. in. are normal, with hardness up to 380 Brinell, although elongation is limited to less than 4 per cent. This can be improved by sacrificing the very high ultimate strengths and hardness by modified cold working and heat treatment, producing very impressive strengths of 60 to 80 tons/sq. in. 295 to 330 Brinell and 5 to 10 per cent elongation.

It is chiefly produced in the form of strip and wire, but is also obtainable as tube and rod. Castings account for a considerable proportion of the annual tonnage, especially in the form of hammer heads, spanners, boulsters, etc.

As a precipitation (age) hardening alloy it is spectacular, but can be rendered soft and malleable by annealing for 2 to 3 hours at 1500° C. and quenching in water.

Welding Procedure

Oxygen and Acetylene: Beryllium coppers cannot be satisfactorily welded by this means.

Electric Arc: Beryllium copper can be welded by shielded Metal Arc Welding using covered 2.5 per cent beryllium copper electrodes, with 1.1 per cent of nickel included. These electrodes operate on D.C. reverse polarity. Manufacturers' instructions should be followed.

Edges should be vee'd out to 90° and a root opening allowed to ensure 100 per cent penetration. A copper or carbon backing plate should be used to prevent burning and collapse.

On large masses a 320° to 370° pre-heat should be maintained.

Downhand welding will be found to be both easier and to produce a better weld if the work can be so positioned.

Otherwise the technique is much the same as for aluminium bronze, but since the metal transfer of the beryllium copper electrode will be found to be globular and erratic, a longer arc should be used. In welding in a groove, care must be taken to draw the arc out when approaching the groove walls to prevent undercutting. Slag must be thoroughly removed between beads.

If maximum hardness of the weld is required, this can be developed by heating to 790° C. and holding at this heat for 1 hour per inch of thickness, quench in water and draw at 320° C. for 1 hour per inch of thickness. The Brinell hardness number should then be about 300.

Fumes when welding and finely divided particles of beryllium copper, such as filings and grindings can be extremely toxic, and wounds caused by this metal can be dangerous. Therefore exercise care when working with it.

28   COPPER NICKEL ALLOYS

Uses

These vary widely according to composition. Because of their high resistance to corrosion by sea-water they are used mainly for marine applications, but they are also widely used for coinage and bullet jackets. The higher nickel content alloys have their main uses as resistances in electrical instruments, but Monel is recognised as probably the best of all materials for steam valves and chemical engineering plants, whilst an addition of iron or cobalt can render them very strongly magnetic.

Composition

Some of these alloys have so little copper as to almost exclude them as copper-based alloys, with the nickel content ranging from as low as 3 per cent to as high as 68 per cent, whilst 1.25 per cent each of iron and manganese is included in Monel, which may also include 2.8 per cent aluminium and 0.3 per cent silicon to promote precipitation hardening.

Properties

The tensile strength varies according to composition from 22 tons/sq. in. to 45 tons/sq. in. with elongation varying from as low as 2 per cent $4\sqrt{area}$ to 35 per cent $4\sqrt{area}$ depending upon the treatment received. Brinell hardness varies from as low as 70 in the annealed condition to 200 in the cold work hardened condition.

They are strongly resistant to corrosion by sea-water; and copper salts derived from the material adjacent to sea-water renders the immediate vicinity poisonous to marine organisms and vegetation, so preventing them from breeding on the metal and causing deterioration.

Welding Procedure

Oxy-Acetylene: Satisfactory welds can be obtained by this method, but rods must be of the same composition and contain enough deoxidiser (manganese or silicon) to protect the metal during welding. The flame should be slightly reducing, and flux specially designed for this alloy is essential to prevent nickel oxide formation, and porosity.

The pool must not be agitated, and once started the weld must be continuous until finished.

Electric Arc: A heavy covered electrode is essential for quality welds by this process. The core will be a 70 per cent copper, 30 per cent nickel wire, and its use will be limited to the jointing of copper–nickel alloys with a nickel content of less than that of the rod. They operate only on D.C. reverse polarity, and work should be done in the flat position.

Correct cleaning and preparation is essential, and all slag must be thoroughly removed before laying subsequent beads.

Moderate peening improves the weld, but neither pre-heat nor post-heat treatment is necessary.

The T.I.G. and M.I.G. processes may be used for welding the alloy using special filler wires. Argon gas shield is used for both processes and for T.I.G. welding D.C. current electrode negative.

29   THE BRASSES

The industrial brasses cover a wide range of alloys of copper and zinc, with or without the addition of other elements such as tin, lead, iron, manganese, nickel, aluminium, and silicon. The range runs from alloys that are almost pure copper to those containing as much as 50 per cent of zinc, and the properties vary just as widely, up to those high-strength brasses with as much as 50 tons/sq. in. tensile as cast.

For commercial purposes the straight copper–zinc brasses, which form the basis of the entire series, are divided into three main groups known as the alpha, alpha plus beta, and beta brasses, previously described under Zinc in the section on elements for alloying with copper.

Cap copper, the gilding metals, clock and engraving brass, are all familiar names, but are of interest only in limited spheres of industry, and will not be considered here, except to mention that cap copper is extensively used for small deep pressings, due to its softness and ductility, whilst the so-called bronze used for decorating shop-fronts, as mullions and trim for plate-glass windows is, more than likely in fact, gilding metal processed to look like bronze.

Whereas the alpha brasses can be worked cold with ease, the beta brasses cannot be deformed to any great extent when cold, without fracture, but at about 500° C. they become very plastic indeed, and should therefore be considered as hot working materials, although in their straight-copper-zinc form they have little industrial use except as brazing solders, the melting point of 50–50 copper–zinc alloy being comparatively low at 870° C., and as most alpha alloys eutectics are above this, the beta alloys can be used to braze the alpha alloys.

The straight brasses consist of copper and zinc, but additions of other elements are frequent, and such metals are often referred to as alloy brasses, and may contain one or more of the following: manganese, nickel, tin, iron, silicon, lead, and aluminium in varying proportions, but either singly or collectively rarely exceeds 5.0 per cent.

The welding characteristics for all brasses in their respective classes are similar, but as electrical and thermal conductivities vary according to composition some adjustment is necessary in techniques.

As they are copper-based alloys with consequent higher thermal and electrical conductivity, higher co-efficients of expansion and lower melting points, it is important to allow greater root openings with larger groove angles, to tack more frequently, using higher pre-heats and interpass temperatures and higher currents for electrode size and work thickness.

In brasses with a high zinc content the zinc will volatilise with high-welding temperature leading to porosity. Brasses with a zinc content above 20 per cent are susceptible to stress-corrosion cracking in service; it is advisable to stress relieve the component for 1 hour at 250° C preferably before welding. Lead in the alloy impairs the weldability of the brass.

30   CARTRIDGE BRASS

Uses

This material is almost unsurpassed for cold deformation by presses and spinning, and is used for producing cartridge- and shell-cases, electric lamp caps, headlamp reflectors and a greater number of other cup- or bowl-shaped articles.

Composition

Nominally, 70 per cent of copper and 30 per cent of zinc, but small variations may be desirable, depending on the qualities required.

Properties

In the fully annealed condition, cartridge brass has a greater elongation under tensile test than any other of the alpha brasses, and this, combined with its tensile strength of over 20 tons/sq. in., makes it a very versatile material.

Like all the alpha brasses, it work hardens when deformed cold, and it must be annealed frequently if it is not to fracture under successive operations. A temperature of 600°C. is usually satisfactory, but care must be taken to prevent grain growth resulting in an orange-peel effect on the surface.

Prone to 'season cracking', this can be prevented by low-temperature heat treatment for 30 minutes at 250°C. Atmospheres containing even traces of ammonia are detrimental to this alloy.

Welding Procedure

Oxygen and Acetylene: Cartridge brass is readily welded by this method and all positional welding is possible. Thorough cleaning is essential, and for thick sections veeing to 90° is necessary. The use of a correct flux is important to control the highly volatile zinc, and the rod should be brass and not bronze. Pre-heating and supplementary heating may be necessary on smaller work, but will be essential on large masses.

Electric Arc: Generally all the brasses weld well by shielded metallic arc using phosphor bronze or aluminium bronze electrodes, which should be the largest practicable with D.C. reverse polarity. To avoid burning out the zinc it is essential to maintain a close arc and a small puddle, and the deposit is best achieved by weaving from side to side over about three times the width of the electrode, dwelling at the side walls to fill in undercut which is prone to happen in this alloy.

All welding by electric arc should be done if possible with this metal in the flat position. Usually no post-heat treatment is necessary.

31   ADMIRALTY BRASS

Uses

For many years this was the only really satisfactory material for ship condenser tubes, but is gradually being superseded by new and better materials more suited to the exacting requirements of marine condensers. It is, however, still widely used for tubes and other parts of condensers using fresh water for cooling, and for castings.

Composition

70 per cent copper, 29 per cent zinc, 1 per cent tin. This is an Alpha Brass.

Properties

Very similar to cartridge brass, except that the 1 per cent tin content improves its resistance to corrosion from salt-water.

Welding Procedure

As for cartidge brass by both oxy-acetylene and shielded metallic arc.

32   NAVAL BRASS

Uses

For structural applications and forgings where corrosion from sea-water is likely. It is obtainable in wrought form such as plate, and as extruded sections of rod, bar, strip, etc., for machining and hot forging. It is a useful material for both sand- and die-casting.

Composition

60 per cent copper, 39 per cent zinc and 1 per cent tin. This is an Alpha plus Beta Brass.

Properties

Tensile strength = 24 tons/sq. in. with an elongation of 40 per cent after hot working between 600° C. and 800° C.; this is a tough and ductile alloy.

Welding Procedure

As for cartridge brass, except that pre-heat temperature should be between 260° C. and 370° C.

33   ALUMINIUM BRASS

Uses

Mainly for condenser tubes for marine applications.

Composition

76 per cent copper, 22 per cent zinc and 2 per cent aluminium, with a small addition of arsenic. When added to brass, aluminium has the same effect on the alloy as would six times its weight of zinc. This alloy falls within the Alpha phase.

Properties

Although the mechanical properties of this alloy are almost identical with those of cartridge brass, the aluminium content confers upon it a protective coating of aluminium oxide, which in the case of damage by abrasion is self-healing, and consequently this metal is extremely resistant to corrosive attack by sea-water.

Welding Procedure

As for cartridge brass.

34   60–40 BRASS (MUNTZ METAL)

Uses

For sheathing small craft, end plates for condensers, architectural trim, large nuts and bolts, hot forgings and castings.

Composition

60 per cent copper, 40 per cent zinc. This is an Alpha plus Beta alloy.

Properties

This is essentially a hot working material and by this means is formed into sheet, plate, rod, and other extruded sections in a great variety of shapes. Hot forgings of all sizes are produced from rod and bar stock, and it is much used for permanent mould and sand castings. Its tensile strength is 23 tons/sq. in. with an elongation of 40 per cent and because of these properties makes an excellent brazing alloy for jointing steel.

60–40 brass with from 0.5 to 3.5 per cent lead included is known as leaded 60–40 brass or turning brass, and is the chief material fed to bar automatics and similar machines, the lead content promoting excellent machinability. Leaded 60–40 brass is also used extensively for hot pressed or forged articles requiring subsequent machining.

The lead content has no marked effect on the tensile strength, but does impair the ductility and impact value, but as it is a hot working material this is of little significance.

Welding Procedure

As for cartridge brass, except that pre-heat and interpass temperature should be between 260° C. and 370° C. The leaded brasses cannot be welded satisfactorily.

35 PRECIPITATION HARDENING ALPHA BRASS

Uses

For machine components, such as gears, instrument pinions, and intricate pressings where strength and hardness are of first importance.

Composition

Variable, but the best known are those containing nickel and aluminium in a matrix of cartridge brass. One proprietary alloy is composed of 72 per cent copper, 20.5 per cent zinc, 6 per cent nickel and 1.5 per cent aluminium.

Properties

Brasses as a rule do not respond to any heat treatment designed to increase strength and hardness, but these comparatively new alloys do. Quenching at 850° C. renders them soft and ductile and they can be rolled, drawn, or formed with the ease of cartridge brass, but re-heating to 500° C. results in considerable improvement with mechanical properties to over 45 tons/sq. in., depending on the degree of cold working between the two heat treatments.

Welding Procedure

Not normally applicable, but as for cartridge brass, although the weld area will never be as strong or as hard as the parent metal, even after appropriate heat treatment.

36 ALPHA+BETA HIGH-TENSILE BRASSES

Uses

These alloys are used for marine propellers and rudders, water turbine runners, the bodies of large rotary or reciprocating pumps, autoclaves, valves, gear-wheels, pinions, axle boxes, non-sparking tools for gas, oil, and explosive industries, non-magnetic mountings for ships; compasses and innumerable other items.

Composition

The basic material of these alloys is 60–40 copper–zinc, but the actual ratio of these two major constituents varies considerably in accordance with those others which are added, for each addition has its own specific effect on the structure of the matrix.

These additions are aluminium, iron, manganese, nickel, lead, tin, and silicon, and jointly or severally have the effect of improving the mechanical properties, or, as in the case of lead, the machineability.

Properties

All engineers are familar with the name Manganese Bronze, which is in fact not a bronze; nor is the manganese content large enough to be regarded as a major constituent, either in volume or its overall effect on the alloy. The manganese bronzes are in reality the alpha plus beta high-tensile brasses.

A sand-casting composition of 54.1 per cent Cu, 39.1 per cent Zn, 1.5 per cent Al, 1.3 per cent Ma, 1.6 per cent Fe and 2.4 per cent Ni will produce a yield stress of 22 tons/sq. in., a tensile strength of 39 tons/sq. in. with an elongation of 19 per cent, whereas a different composition for chill casting would produce 21 tons/sq. in. yield stress, 42 tons/sq. in. tensile strength and 22 per cent elongation, whilst various compositions for hot-worked material will produce yield stresses from 10 to 25 tons/sq. in., tensile strengths from 30 to 44 tons/sq. in., and elongations ranging from 20 to 43 per cent. Thus it will be seen that there is great latitude in composition, and equal latitude is allowed to the manufacturer, just so long as the required mechanical properties are met.

The high-tensile brasses are used in the form of castings, and after hot working by forging or extrusion, where strength and toughness and good resistance to corrosion are of primary importance, but under certain conditions a form of attack known as 'de-zincification' occurs, in which the zinc is destroyed by insidious penetration of a corrosive agent, and other alloys of a more homogenous nature are to be recommended in preference to the high tensile brasses where corrosive conditions are particularly severe. Whilst it is safe to soft solder these alloys in their unstressed condition, they will absorb solder in their stressed state and crack on cooling.

Welding Procedure

As for cartridge brass, except that pre-heat temperatures of 260° C. to 360° C. should be used, and great care taken to maintain as small a puddle as possible to prevent zinc volatilisation.

37 THE BRONZES AND GUNMETALS

As has been indicated the true bronzes are alloys of copper and tin, but additional elements which give their particular alloys better characteristics for different usage, are often included. These are mainly zinc, phosphorus, nickel and lead, and those bronzes that contain zinc are usually classed as gunmetals.

As usual the bronzes fall into the two main classes of the wrought alloys such as rod, wire, and strip, and the more complex alloys used for castings.

Normally, 8 per cent of tin is the highest amount present in bronzes and phosphor-bronzes for sheet, strip, and wire, although the solid solubility is almost twice as high at 550° C., and, by suitable heat treatment, bronzes relatively high in tin content can be brought into an homogenised condition suitable for cold working into wrought products, and these have added resistance to corrosion, and high elasticity.

Although not Bronzes, the Stellites and Manganin have been included in this section for convenience.

In addition to those alloys listed, there are of course many others such as bell metal, speculum metal, antimonial bronze and nickel silvers, all of which have their rightful place in industry, but which are of little interest as engineering materials.

There are too, the Copper–Manganese–Aluminium Alloys known as the Heusler alloys, which although containing no iron, can be made very strongly and permanently magnetic; and the Duplex Aluminium Bronzes which are of a very complex nature but highly suitable for gravity die casting.

Sintered bronze is another widely used copper based product which is made by copper and tin powders in correct proportions, being placed under pressure in a former into a heated and reducing atmosphere and so fused into a solid of the required shape. Various degrees of porosity can be produced, and when impregnated with oil make excellent pre-lubricated bearings. Sometimes graphite, an excellent lubricant itself, is introduced at the sintering stage. Another use for this process is for making filters for the oil and chemical industries.

There is virtually no end to the useful compositions of bronze alloys that can be produced, but those mentioned are of most general interest to engineers.

38   PHOSPHOR BRONZE

Uses

The phosphor bronzes are highly favoured for spring making and wire products in the wrought form, whereas cast phosphor bronze is extensively used for bearing applications and general castings as in gear blanks, valves for high-pressure steam systems, pump bodies and impellers, etc.

Compositions

Phosphor bronzes may be roughly divided into three main groups:

1. Bronzes of straight copper–tin in which phosphorus has been used in balance only as a de-oxidiser.

2. Bronzes with up to 9 per cent tin and only a trace of phosphorus, for wrought forms of bar, sheet, strip, tube, and wire.

3. Bronzes with more than 9 per cent tin and phosphorus ranging from 0.2 to 2.0 per cent maximum, for castings.

A small amount of nickel may be included as this improves the pour, and helps in producing sound castings, refines the grain and bestows good machining properties.

Properties

Chill-cast and spun-cast are superior in structure and texture to sand-cast phosphor bronze, being hard, tough, and abrasion resistant.

Although the maximum tensile and compressive stresses are comparatively low at 23 and 15 tons/sq. in. respectively, the Brinell hardness can be as high as 160 with a 10 m/m ball loaded to 3000 kg. for 30 seconds. For this reason it should not be used as bearings on mild steel shafts, but only on shafts of steels with higher Brinell hardness numbers than the phosphor bronze.

This alloy machines well, and is easily cold worked, but because of its very narrow plastic range must be hot worked at between 620°C. and 665°C. In highly stressed conditions the phosphor bronzes suffer from stress corrosion, and severely worked parts should be stress relieved before being put into service, but care must be taken to avoid thermal shock by the sudden application of heat, as they are also subject to fire cracking.

Annealing temperatures vary between 490°C. and 760°C., depending on the composition.

Welding Procedure

The oxy-acetylene welding process is not ordinarily now employed in welding the copper–tin alloys, because the high thermal conductivity spreads the heat, causing irregular expansion and contraction which results in cracking of this hot short material, but when this process must be used, 1.5 per cent tin rods should be used as a filler, or 8 per cent tin rods for hard surfaces. A good brazing flux is essential, and a neutral flame is preferred. A narrow heat zone must be maintained for reasons stated above, and weaving of the flame must not be allowed.

Jointing by the oxy-acetylene method will be found to be much easier by conventional bronze welding with copper–zinc rods than by welding with copper–tin rods, and the former should always be preferred except where homogeneity is essential. Pre-heating is required for both methods.

Electric arc welding by shielded metallic arc is the most satisfactory means, but the tin–bronzes tend to flow sluggishly, and so require pre-heat to temperatures of 150° C. to 200° C., especially on heavy sections. A backing strip is desirable and all welding should be done in the flat position, although most electrodes can be used in vertical and overhead passes if necessary. Normally

D.C. reverse polarity is preferred, but electrodes for A.C. operation are available.

Tin–bronzes inherently have a coarse grain structure which is reproduced in weld deposits and resulting in low mechanical properties unless proper precautions are taken. This coarse grain structure can be refined to some extent by hot peening of the deposit.

39 GUNMETAL

Uses

Largely used for naval purposes and for valves and fittings on steam-raising plant, ornamental bronze work, plaques, and slip-rings or collector-rings on electrical rotating machinery.

Composition

Admiralty Gunmental consists of 88 per cent copper, 10 per cent tin, and 2 per cent zinc, and its American equivalent 88 per cent copper, 8 per cent tin and 4 per cent zinc. Their properties are almost identical.

Properties

Originally named because of their suitability for making naval guns, the composition was 90 per cent copper and 10 per cent tin, as this gave the strength, toughness, elasticity, and ability that were needed to resist shock. The inclusion of zinc now contributes to high-corrosion resistance in adverse atmospheres. Tensile strengths of 20 tons/sq. in. are normal with over 20 per cent elongation, and hardness values up to 95 Brinell. They have good hot and cold working properties, and anneal between 480° C. and 760° C., but care must be taken to avoid thermal shock, as they are prone to fire cracking and stress corrosion. Severely worked parts should be stress relieved before being put into service.

Welding Procedure

As for phosphor bronze, except that those containing lead cannot be satisfactorily welded.

40 THE LEADED BRONZES

Uses

Marine castings, pump impellers and bodies, valve bodies, gears, bearings.

Composition

70 to 85 per cent copper, 5 to 10 per cent tin, 5 to 25 per cent lead.

Properties

Depending on the composition of the alloy the tensile strength varies between 9 and 14 tons/sq. in. with the higher lead contents producing the lowest strengths.

Variations with small quantities of tin or very rarely silver, and up to 30 per cent of lead, because of their higher pressure per unit area have been developed for high-duty bearings for diesel and aircraft engines.

The leaded bronzes with tin up to 5 per cent are good general-utility metals with reasonable strength, good corrosion resistance, foundry properties and machining characteristics. One such alloy is known popularly as 85/5/5/5, being composed of copper, tin, lead, and zinc in these proportions.

The high lead–tin bronzes such as the 80/10/10 alloy have excellent anti-friction and machining properties, with high resistance to wear, the lead content itself having inherent lubricating properties.

Welding Procedures

Oxy-Acetylene: Welding is not satisfactory by this means, and brazing is to be preferred, but in the high-lead-content alloys even this method will produce very poor results, due to the lower melting point of the lead, which does not in fact alloy with the copper, but exists in the metal as very small droplets, which begin to melt at 621° F. and sweat over the job. This prevents fusion and causes voids in the work, where the lead has melted out.

41 NICKEL BRONZES

Uses

The most important applications of these alloys is in the production of valves, valve gear, and the wearing parts of pumps, impellers, shafts, etc., on steam-raising plant and boiler feed equipment.

Composition

In the 88/10/2, 88/8/4, and 85/5/5/5 gunmetals, the inclusion of up to 1.5 per cent of nickel improves the mechanical properties, pressure tightness and density of castings, and helps to prevent the segregation of the lead in heavily leaded cast-bearing bronzes.

High-content Nickel Bronzes may contain from 15 to 60 per cent nickel for special purposes, but a more normal content ranges from 3 to 5 per cent nickel. From these alloys lead must be rigorously excluded, as it severely interferes with heat treatment, to which the nickel bronzes respond perfectly.

Properties

With appropriate heat-treatment strengths of over 30 tons/sq. in. can be achieved with hardness values up to 145 Brinell, but elongation is usually low at about 3 per cent. These alloys retain their strength at elevated temperatures and are highly resistant to mechanical wear and corrosion, and erosion by water and steam. They have excellent cold and hot working properties, except for hot forging because of poor flow qualities. They are very resistant to stress corrosion and corrosion fatigue.

Welding Procedure

As for phosphor bronze for both gas and arc methods, except that the leaded qualities cannot be successfully welded, and that the filler must be of the same composition as the parent metal and contain manganese or silicon in sufficient amounts to act as a de-oxidiser to protect the metal during welding. The flame adjustment should be slightly reducing. The weld once started must be continued to the end without stopping.

42   SILICON BRONZES

Uses

In both the wrought and cast condition these alloys find their main uses in chemical engineering applications, especially for storage tanks, piping systems, pickling crates. They also have use as pump parts, gears, shafts, nails for boat-making, and other important uses in the paper-making industry. Their high electrical resistance also makes them suitable for resistance welding.

Composition

Varies widely as to the copper and tin content but in the wrought product 3 per cent of silicon is added, whilst for casting 4 to 5 per cent is added, with small additions of iron, zinc and up to 1 per cent of manganese.

Properties

These are high-strength materials, cold working raising the tons/sq. in. from 23 in the annealed condition to over 50. The addition of silicon not only enhances the strength but confers resistance to corrosion by acids, and improves the weldability. Their work-hardening properties are good, and annealing is achieved at temperatures between 490° C. and 760° C. but they are subject to fire cracking and stress corrosion failures.

Welding Procedure

Oxy-Acetylene: Because of the silicon content, these alloys are very readily welded by this process, using a filler rod of the same composition, and a flux of high boric acid content. A slightly oxydising flame should be used to help maintain as small a weld pool as is consistent with the work. Brazing is also easily effected with copper–zinc rods where dissimilar metals are not objectionable.

Electric Arc: The shielded metal arc process is also highly effective in jointing silicon bronze, using either silicon bronze or aluminium bronze electrodes.

The character of this alloy facilitates welding and no pre-heat is necessary, but rapid welding is necessary, keeping the puddle small with a short arc to prevent over-heating. Downhand welding is preferred if the work can be so positioned. Moderate peening will improve the mechanical property of the joint, but no post-heat treatment should be used.

43   STELLITES

Uses

Hard facing for rolls, hot cutting edges, bucket lips, digger teeth and machine and tool bits, hand-cutting tools, valve seats and exhaust valves.

Composition

Cobalt 45 to 55 per cent, chromium 30 to 35 per cent, tungsten (wolfram) 12 to 17 per cent, carbon 1.5 to 2.0 per cent.

Properties

The Stellites are not steels, as is often believed, and they contain no iron. Their properties are similar to those of high-speed steels, in that they are extremely hard, and retain this hardness at high operating temperatures, and even at red heats. They are extremely resistant to corrosion from acids and atmosphere. In addition they polish well and in that condition have high reflective powers, and are difficult to scratch. Stellites cannot be produced in wrought forms, and as the plastic range is extremely fine they cannot even be worked hot. They are therefore always cast and ground to size and shape.

No heat treatment is necessary.

Welding Procedure (Deposit)

Oxy-Acetylene: The techniques for depositing surface protection metals is different from normal fusion welding, in that inter-alloying must be avoided. This can be achieved by using a flame with an excess of acetylene so that the surface absorbs sufficient carbon to enable it to melt at a lower temperature than the metal immediately below the surface.

The excess acetylene flame should be $2\frac{1}{2}$ times as long as the normal neutral cone, and kept soft. Keep the inner cone almost in contact with the surface metal and apply the rod between the inner cone and the hot base metal. This carburising flame will give the red hot base metal the appearance of sweating, and as the majority of the heat is directed on the rod, this will melt, fall on to the sweating base metal, which if at the right temperature will cause the stellite to spread uniformly.

This sweating technique can only be employed on iron- and cobalt-based deposits, and when the base material is steel.

The base metal must be thoroughly descaled and cleaned.

Electric Arc: Depositing by shielded metallic arc welding is satisfactory and is done in much the same manner as conventional electric arc welding, except that, to prevent dilution, the deposit should be applied with as low heat as possible.

Current, voltage, polarity, and other conditions are specified by the electrode manufacturers and these should be closely adhered to.

Stellite may also be applied as an overlay by the submerged arc process.

44 COPPER–MANGANESE–NICKEL ALLOYS

Uses

This is not a bronze as no tin is included. It is popularly known as manganin, and used for fixed resistances on electrical instruments.

Composition

Copper 84 per cent, manganese 12 per cent, nickel 4 per cent, but 60 per cent copper, 20 per cent manganese and 20 per cent nickel alloys produce excellent precipitation hardening characteristics.

Properties

Tensile strength 24 to 85 tons/sq. in. with hardness values from 45 to 420 Brinell, but with extremely limited elongation.

Welding Procedure

Not applicable.

Further Reading

Copper and its Alloys in Engineering and Technology. Aluminium Bronze. The Welding of Copper and its Alloys. Copper Development Association, 55 South Audley Street, London W.1.

Materials for Engineering Production. P. S. Houghton, A.M.I.MECH.E. The Machinery Publishing Company Ltd., Brighton, Sussex.

Welding Handbook. American Welding Society, 33 West 39th Street, New York 18, N.Y., U.S.A.

Metals in the Service of Man. William Alexander & Arthur Street. Pelican Books.

TEMPERING COLOUR CHART

(For carbon tool steel only)

° C.	Colour Change	° F.
330	Grey	626
310	Pale blue	590
300	Medium blue	572
295	Full blue	563
290		554
285	Purple blue	545
280		536
275	Purple	527
270		518
265	Red-brown	509
260	Brown-yellow	500
255	Orange	491
250	Light orange	482
245		473
240	Deep straw	464
235		455
230	Straw	446
225		437
220	Light straw	428

Applications (from highest to lowest temperature):

- Springs
- Wood saws
- Circular saws for metal
- Screw-drivers
- Cold chisels for wrought iron
- Moulding and planing cutters (soft wood). Needles
- Cold chisels for cast iron and firmer chisels
- Cold chisels and setts for steel
- Hot setts
- Axes and adzes
- Press tools
- Augers
- Flat drills for brass
- Twist drills and cooper's tools. Stone cutters
- Wood borers, plane irons and reamers
- Hard wood planing and moulding cutters, punches, dies, cups, snaps, gouges
- Chasers, shear blades for hard metals
- Taps, mill chisels and picks
- Screw-cutting dies
- Rock drills
- Boring cutters
- Leather punches and dies
- Milling cutters
- Bone and ivory cutting tools
- Wood-engraving tools, planing and slotting tools for iron
- Paper cutters
- Planing and slotting tools for steel
- Hammer faces, mint dies
- Screwing dies for brass
- Light turning tools, steel engraving tools
- Lathe tools, scrapers and brass turning tools

This chart may be used as a guide to tempering temperatures. In almost every case, hardening of tools demands tempering which relieves stresses and induces toughness to a degree suitable for the work on which the tool will be employed.

In those tools where the maximum hardness is required, they are made much tougher and less liable to fracture, without appreciable loss in hardness, if they are tempered at 100° C. to 130° C.

The temperature is estimated by the surface colours of the steel, formed by oxidation, but these will only show if the steel is free from scale. It should, therefore, be cleaned with emery cloth or on the grindstone, but must not be polished, glazed, or burnished as this results in different colour changes.

HARDENING TEMPERATURE CHART

°C.	Colour Change	°F.
565	Dull red	1049
600	Dark red	1112
700	Blood red	1292
730	Dull cherry red	1346
740	Low cherry red	1364
750	Med. cherry red	1382
760	Med. cherry red	1400
770	Cherry red	1418
780	Cherry red	1436
800	Cherry red	1472
825	Bright cherry red	1517
850	Full red	1562
900	Bright red	1652
920	Bright red	1688
955	Full bright red	1751
980	Orange	1796
1150	Yellow	2102
1250	White	2282
1300	Bright white	2372
1330	Bright white	2426
1350	Incandescent white	2462
1360	Incandescent white	2480

Note: These colours must be viewed in shade and not in direct light, as this would cause a dulling effect indicating lower temperatures.

Degrees Centigrade $\times \frac{9}{5} + 32 =$ Degrees Fahrenheit.

Degrees Fahrenheit minus $32 \times \frac{5}{9} =$ Degrees Centigrade.

This chart of temperature colours ranges from visible red to scintillating incandescent white and may be used as a guide to heats when hardening carbon or tool steel. The following points should be observed:

1. Do not heat the steel too quickly as distortion will occur and the steel might burst on quenching
2. The steel must be heated uniformly and through and through otherwise it might crack on quenching
3. Do not over-heat the steel as this will cause burning and it will be ruined by oxidation
4. Do not over-soak the steel as this will result in decarburisation of the surface, causing hard and soft places when quenched
5. Do not quench at temperatures higher than recommended by the makers; this will cause severe distortion

GLOSSARY

Abrasion: Wearing away.

Age Hardening: A change in the physical properties of a material, apparently occasioned only by the passage of time.

Alloy: A mixture of several metals.

Anneal: To soften a metal by heat treatment, by heating to the critical temperature, holding that heat until uniform throughout the mass, and cooling slowly. This makes the steel more amenable to machining, stamping, forming, etc., and/or alters its properties for some specific duty.

Argon Arc Welding: Arc welding process in which the weld metal is protected from contamination by other elements in the atmosphere, by a shield or envelope of argon gas liberated from the cover of the electrode or introduced to the weld area by other means.

Argon Gas: Inert gas used in welding and filling electric lamp envelopes.

Atomic Hydrogen Welding: Arc welding process in which the electric arc is drawn between tungsten electrodes positioned in a jet of hydrogen gas.

Austenitic: (1) A steel in which a solid solution of carbon or iron carbide exists; (2) A steel containing sufficient nickel, nickel and chromium, or manganese, to enable the retention of austenite at atmospheric temperature. Steel with cold-hardening properties.

Basic Steel: A steel made by the basic process and low in hydrogen. Acid steels are high in hydrogen ions and unsuitable for certain applications.

Binary Alloy: An alloy consisting of only two components.

Bloom: A half-finished rolled or forged section of steel greater in cross-sectional area than 25 sq. in. A billet is less than 25 sq. in. cross-sectional area.

Bourdon Tube: A semi-circular tube of oval cross-section which tends to straighten as the pressure inside increases. Used extensively in the manufacture of pressure-gauges.

Bright Drawn: A final operation in steel finishing in which the bar is drawn cold through a die smaller in size than the steel. This removes the carburised black surface (mill scale), leaving a high bright finish and at the same time finishing the steel to precise dimensions.

Brinell Hardness: A test applied to steel to determine its hardness rating, in which a hard steel ball 10 mm. in diameter is loaded on a steel sample with a predetermined weight for 15 seconds, and the diameter of the indentation measured, from which is calculated the Brinell Hardness number (see DP Hardness and Rockwell Hardness).

Brittle Zone: A zone in steel where heat has been applied to a temperature high enough, and cooled quickly enough, to cause local hardening varying from the hardness of the bulk of the metal and liable to cause fracture.

Carbon: C. Element. Occurs in various forms, but in steel as cementite, and is combined with the steel. In cast iron it is not combined, is known as free carbon, and is in graphite form.

Carbon Arc Welding: Arc welding process in which one electrode is a carbon rod, and the other is the work, the filler metal being supplied by a separate rod, held in the arc. Used mainly for filling up holes and building up metal.

Carbon Monoxide: CO. A poisonous gas, the properties of which render it useful as a reducing agent.

Carburising: The process of carbon absorption by steel under heat, from contact with carbonaceous materials either solid (leather or charcoal), liquid (molten cyanide salts or mineral oils), or gas (ammonia), to enable hardening to take place.

Case Hardening: Surface hardening of steel by carburising, cyaniding, or nitriding. Such hardening is only a few thousandths of an inch deep.

Casting: A mass of metal shaped by pouring it in a molten state into a mould.

Cementite: Iron carbide. Constituent of iron and steel.

Chill Cast: A process of casting molten metal into a cold metal mould which accelerates cooling and so gives great hardness and density to the whole or part of the casting.

Chromates: The salts of chromium trioxide. The process of chromating magnesium castings protects the surface metal from attack by the atmosphere, which otherwise would cause the metal to decay.

Chromatic Heat Indicators: Coloured crayons, the marks from which change to specific other known colours under the influence of certain temperatures between 65° and 670°. (Chroma—Greek—colour.)

Chromium: Cr. Metallic element.

CO_2 Arc Welding: Fast welding process specially suited to light gauge sheet and plate fabrication, in which carbon dioxide gas is used to shroud the arc, making the weld deposit virtually slag-free.

Cobalt: Co. Magnetic metal. It improves the cutting ability of high-speed tools and alters the magnetic properties. Also used to make blue stained glass.

Co-efficient of Expansion: The expansion of a material per unit of its length for each degree of rise in temperature. For steel this is roughly .000006 per degree of Fahrenheit.

Cold Hardening: (Work hardening—age hardening.) The hardening of metals by virtue of the change in their structure brought about by beating, drawing, hammering, bending, vibration, etc., and often occurring as a result of the work the metal does, sometimes resulting in fatigue and failure.

Cold Short: An undesirable property in steel or wrought iron as a result of the inclusion of too much phosphorus, causing a tendency to fracture when cold.

Columbium: (Niobium.) Rare grey metal included in welding quality alloy steel to give stability, ensuring that after melting in laying the bead, it solidifies with its original characteristics.

Complex Alloy: An alloy consisting of more than two components and presenting an involved matrix.

Corrosion: Chemical action on the surface of materials, especially metals, by the action of air, moisture, and chemicals.

Corrosion Fatigue: The failure of a metal when subject to repeated cycles of stress while exposed to corrosive attack. Even relatively pure water will lower the endurance limit considerably.

Creep: The slow continuous deformation of a metal at constant stress exhibited in steel, nickel, copper and copper alloys at elevated temperatures, and in lead, tin, and zinc at room temperatures.

Critical Temperature: That temperature at which in plain carbon steel, on heating or cooling, a change takes place in its molecular structure and at which point the steel continues to absorb or shed heat without increase or decrease in temperature. These critical points have a direct relation to the hardening of steel and unless temperatures reached are high enough to change the pearlite into austenite no hardening can take place on quenching. Similarly, in normalising, unless the critical temperature is reached after which cooling slowly in air takes place, the austenite will not revert to pearlite and softening will not take place. Steel loses its magnetic properties at these heats, hence magnetic indicators can be used to show when they have been reached.

Cyanide Hardening: A process of introducing carbon and nitrogen into the surface skin of mild steel by holding it submersed in molten cyanide salts.

De-oxidising: The process of elimination of oxygen from molten metals by adding elements with a high oxygen affinity, which form oxides that rise to the surface and so can be skimmed off.

De-zincification: A process of the decay of the zinc in zinc-bearing alloys, by corrosive attack, resulting in a serious weakening of the metal.

DP Hardness: Diamond Pyramid Hardness. A method of testing the hardness of metal surfaces by pressing a diamond pyramid point into it for a predetermined time at a predetermined load and measuring the indentation. There is a close relationship between DPH and Brinell Hardness results.

Ductility: Property of metals which enables them to undergo plastic deformation without breaking.

Elastic Limit: That stress beyond which a metal will not recover to its original length or shape.

Electrolytic Refining: Method of producing pure metals by making the contaminating elements the anode in an electrolytic cell and so depositing a pure cathode.

Elongation: The percentage a metal stretches from its original length at its ultimate tensile strength, i.e. the point of failure.

Endurance Limit: The maximum stress for any material below which fractures do not occur, regardless of how many reversals of stress take place. For steels this limit is between 6 to 10 million cycles of stress, and in mild steel is about three-quarters of the yield stress.

Eutectic: The minimum freezing (solidifying) temperature attainable by mixing two or more substances capable of forming solid solutions and having the property of lowering each other's freezing point.

Fatigue: That point at which a metal component, subjected to repeated reversals of stress, will fail. (Can be corrected before failure, by normalising.)

Fire Cracking: Fine hair cracks appearing in metals (and ovenware) as a result of heating and cooling.

Fusion: The joining of parts of the same material by melting the edges together.

Gassing: A process in which when copper is heated in an atmosphere containing hydrogen, the hydrogen diffuses into the metal and combines with oxygen residues left in the copper. Thus the hydrogen and oxygen form moisture from which the heat produces steam which cannot diffuse out. A high steam pressure will result in diminished cohesion of the metal structure, causing failure. Hence the need to keep copper as free of oxygen as possible.

Grain: The structure of a metal as seen under a microscope.

Heat Treatment: Generally any controlled heating operation done on solid metal with the object of changing its immediate characteristics.

Helium Gas: He. Element. The lightest of the rare gases and used for inert gas welding on D.C. welding plants.

Homogenised: Made to be of uniform composition throughout.

Hot Short: An undesirable property in steel or wrought iron due to an excess of sulphur, causing the metal to be liable to fracture when hot.

Impact Extrusion: The forming of bars, tubes, or sections of intricate shape by forcing the metal, hot or cold, by impact, through dies corresponding to the required shape.

Inert Gas Metallic Arc Welding: Any welding process using an inert gas such as argon, carbon dioxide, hydrogen, etc., for shielding the arc formed by the metal electrode.

Ioniser: Device for improving the electrical conductivity of the inert gas shrouding an arc during welding.

Izod Impact Test: A test in which a rigidly held notched bar is struck by a striker carried on a pendulum. The energy required to fracture the bar is calculated from the height from which the pendulum falls.

Lamination: A thin layer.

M.I.G.: Metal Inert Gas welding (CO_2). Consumable wire electrode fed through gun protected from contamination by gas shield CO_2-argon, etc., dependent on wire.

Matrix: The material of a solid in which the grains forming the solid are embedded. (Cement is the matrix of concrete.) In metallurgy, the matrix of a specific alloy, after having been polished and etched, will present a pattern characteristic of the alloy and the heat treatment it has received. From this pattern a great deal of information about the alloy can be ascertained.

Nitriding Process: A process for producing hard surfaces on special steels by heating in gaseous ammonia.

Nitrogen: N. Odourless, invisible, chemically inactive gas, forming four-fifths of the earth's atmosphere. Undesirable in steel, causing brittleness.

Normalising: The heating of steel to above the critical temperature, followed by slow cooling in air of room temperature. Normalising produces the effect of putting the steel into a uniform unstressed condition, making it less subject to failure.

Plastic Range: The temperature range between which solids become plastic and easily worked. Some solids have a wide range, enabling much work to be done before cooling so solidifies the metal that it cannot be further deformed, whilst others have such a narrow range that the only possible way of moulding them to shape is by casting.

Precipitation Hardening: The phenomenon which results in an increase of hardness with time either at atmospheric or elevated temperatures. This increase is due to a change in the structure of the solid occasioned by the reaction of its composition.

Reducing Atmosphere: An atmosphere that has been manipulated to exclude certain normally present, but undesirable, elements which might adversely affect a process.

Roak: A seam or lamination in steel.

Rockwell Hardness: A method of testing the hardness of metals by measuring the depth of penetration of a conical diamond point (for hard metals) or of a steel ball (for soft metals) under a specified weight for a specified time.

Scale: Black iron oxide enveloping iron or steel which has been hot worked.

Season Cracking: Cracking at stressed points due to age hardening.

Selenium: A non-metallic element that improves the machining qualities of copper and stainless steels, but in most alloys is detrimental. Used also as photo-electric cells, temperature measuring devices and as a colouriser for stained glass.

Shielded Metallic Arc Welding: Any process of welding in which the arc is struck by a metallic electrode and is shrouded by an inert gas, usually generated from the cover of the electrode.

Specific Gravity: The weight per unit volume of a substance compared with the same volume of water at a given temperature.

Spun Cast: Tubes, pipes, and other cylindrical shapes and discs, cast in moulds, rotating at quite high speeds, the molten metal being introduced by a long spout. The process is called centrifugal casting.

Slag: Waste glass-like material derived from the impurities released during the blast-furnace stage of steel making, and being lighter than the molten steel, floats to the surface from where it can be skimmed off.

Stress: A force on a member divided by the area which carries the force, and usually expressed in pounds or tons per sq. in. There are three types of stress: compressive, tensile, and shear.

Stress Corrosion: Corrosion occurring at stressed points in metals resulting in disintegration of the metal at those points.

Stress Relieved: Metal heated to a temperature below the critical point followed by slow cooling to relieve the internal stresses set up by machining, pressing, or other manipulation. Uneven cooling will set up a different combination of internal stresses.

T.I.G.: Tungsten Inert Gas welding. Arc formed between tungsten electrode and work; arc shielded by inert gas usually argon, filler rod normally applied manually.

Tempering: Imparting a definite degree of hardness to steel by carefully controlled heating and quenching.

Tensile Strength: The (pulling) stress which has to be applied to a material to break it, and measured in pounds or tons per sq. in.

Thermal Shock: Shock caused by the sudden application of heat to a metal sensitive to heat by virtue of its great coefficient of thermal expansion. May cause splitting or cracking.

Volute: A conical spiral in which each succeeding turn overlaps the previous turn.

Warp; Permanent distortion of a part from its original form, usually due to the effects of heat.

Yield: That point just beyond the elastic limit at which the material is unable to effect recovery to normal, and at which permanent elongation results.

$4\sqrt{Area}$: Four times the square root of the cross-sectional area. A means of expressing the relationship between the cross-sectional area and the length of a sample undergoing tensile test, in order that the relationship between these two dimensions remains constant, for determining the elongation per cent of the length.

That the multiplier varies from country to country (it is 11.3 in U.S.A.) does not matter, so long as it is understood that the results are achieved on that constant in national use.